Wireless Power Transfer for Electric Vehicle Charging

WIRELESS POWER TRANSFER FOR ELECTRIC VEHICLE CHARGING

Ognjen Grkinic

ARCLER PRESS

www.arclerpress.com

Wireless Power Transfer for Electric Vehicle Charging

Ognjen Grkinic

Arcler Press
4164 Lakeshore Road
Burlington, ON L7L 1A4
Canada
www.arclerpress.com
Email: orders@arclereducation.com

ISBN: 978-1-77956-432-0 (Hardcover)

ABOUT THE AUTHOR

Ognjen Grkinić is a physicist with a profound expertise in applied and computer physics, particularly in data science and plasma physics. He is currently pursuing his PhD at the Faculty of Physics, University of Belgrade, where his research focuses on the behavior of ionized gases and plasma within the Attosecond-Plasmas Lab. Ognjen's academic achievements include a Master's degree in applied and computer physics, where he developed advanced software for processing scientific data, earning a near-perfect GPA of 9.80/10.00. In addition to his research, Ognjen has contributed significantly to education, having taught physics and mathematics at both high school and primary school levels. His passion for teaching is matched by his meticulous approach to scientific inquiry, which is evident in the books and papers he has authored and edited in the field of physics and other natural sciences. Beyond academia, Ognjen has worked as a Multimedia Transcriber at Arcler Education Inc. and as a Data Labeling Quality Specialist at Scale AI, roles that have further refined his analytical and technical skills. Fluent in multiple languages, Ognjen is also an active volunteer, dedicating his time to educational and community initiatives. Ognjen Grkinić's diverse experiences in research, education, and professional work underscore his commitment to advancing the field of physics and contributing to the broader scientific community.

TABLE OF CONTENTS

LIST OF FIGURES

LIST OF TABLES

PREFACE

The integration of Wireless Power Transfer (WPT) technology into the electric vehicle (EV) industry has significantly transformed the way EVs are charged, enhancing convenience, and promoting broader adoption. WPT systems enable the transfer of electrical energy from a power source to an electric vehicle without the need for physical connectors, using electromagnetic fields, typically through inductive coupling. This technology enhances the user experience by eliminating the need for manual plug-in procedures and reducing wear and tear on physical connectors, thereby fostering greater efficiency and reliability in EV charging.

The chapters within this book provide a comprehensive exploration of WPT systems, their fundamental principles, technological advancements, and practical applications in the context of electric vehicle charging. From understanding the basics of WPT to evaluating its regulatory framework and contemplating future prospects, each chapter is meticulously crafted to offer a structured learning experience.

Chapter 1 serves as an introduction to WPT, elucidating its historical background, underlying principles, and various methodologies. Subsequent chapters delve deeper into the classification criteria, associated technologies, and market perspectives of WPT systems. Chapter 3 navigates through the specifics of wireless chargers for electric vehicles, covering topics such as charging modes, standards, and market insights.

Further, the book expounds on the static and dynamic aspects of WPT for EVs in Chapter 4, followed by an exploration of technological advances in electric vehicles in Chapter 5. Chapter 6 provides a focused examination of inductive wireless power charging systems, shedding light on their architecture, current state, challenges, and opportunities. Lastly, Chapter 7 probes into the intricacies of WPT in online electric vehicles, elucidating design considerations, mechanisms, and optimization techniques.

Finally, Chapter 8 wraps up the discussion by evaluating the regulatory landscape, future prospects, and emerging trends in EV wireless power transfer technologies. Each chapter is supplemented with learning objectives, review questions, and references to facilitate a comprehensive understanding of the subject matter.

This book has been written specifically for students and scholars to meet their needs in terms of knowledge and to provide them with a broad understanding of business information systems.

—Author

CHAPTER 1

INTRODUCTION TO WIRELESS POWER TRANSFER

LEARNING OBJECTIVES

At the end of this chapter, readers will be able to:

1. Understand the historical development and significance of Wireless Power Transfer (WPT).
2. Know the categories and principles underlying different WPT technologies.
3. Explore the basic structures and theories used in WPT systems.
4. Know the various methods employed to achieve wireless power transfer.
5. Understand WPT systems and charging mechanisms for Electric Vehicles (EVs).
6. Recognize the diverse applications and potential of WPT technology.

INTRODUCTORY EXAMPLE

One prominent application of Wireless Power Transfer (WPT) is in the automotive industry, particularly for electric vehicles (EVs). Consider the case of Tesla, a leading company in electric vehicle manufacturing. Tesla has been at the forefront of innovation, not only in electric vehicle design but also in charging infrastructure. For instance, Tesla introduced its groundbreaking "Tesla Wireless Charging" technology, aiming to simplify the charging experience for EV owners. With this technology, drivers no longer need to worry about plugging in their vehicles; instead, they can simply park over designated charging pads installed in parking lots, garages, or even on roadways. The system enables the convenient and efficient charging of vehicles by wirelessly transferring power from the charging pad to the onboard batteries, eliminating the need for cords or connectors.

UNIT INTRODUCTION

This chapter provides a comprehensive overview of the technology behind Wireless Power Transfer (WPT). This text offers a comprehensive overview of WPT technology, covering its historical background, different classifications, and various applications. It explains the fundamental structures and theories related to magnetically coupled wireless power transfer (WPT) technology (Zhang et al., 2018).

It delves into different structures and techniques used in wireless power transfer, with a focus on inductive coupling and resonance coupling, which are the primary methods currently used for EV charging. While power transfer through radio waves and optical waves remains a topic of research, these methods are not yet commercially viable for large-scale electric vehicle charging due to efficiency and safety concerns. In this chapter, explores the application of wireless power transfer (WPT) within the realm of electric vehicle charging. It explores the underlying principles and complexities of the charging process. Lastly, the chapter discusses the various uses of wireless power transfer (Li & Mi, 2014).

1.1. OVERVIEW OF WIRELESS POWER TRANSFER

Learning Objectives:

- Understand the concept of Wireless Power Transfer (WPT).
- Differentiate between the categories and principles of WPT technologies.

1.1.1. What Is Wireless Power Transfer

A method called wireless power transfer, sometimes referred to as wireless energy transmission or wireless charging, enables the transfer of electrical energy from a power source to an electrical device without the need for physical cables or connectors. The invention transforms power connections by providing a practical and effective way to power or charge a variety of devices (Patil et al., 2017).

1.1.2. Background and History of WPT

The era of electricity has arrived since the Second Industrial Revolution. Electric equipment is now widely utilized and essential to daily living. Currently, direct connections between metal conductors are the primary method used for the transmission, distribution, and use of electric power.

Numerous concerns arise from the traditional method of power transfer, including safety, feasibility, and convenience in certain applications. Sparks can potentially arise from fatigue, wear, and corrosion of connecting wires, which can have an impact on the longevity and safety of the equipment. Given the fast-paced advancements in technology and the significant enhancements in everyday life, a wide range of household appliances and consumer electronics have become increasingly prevalent (Hui, 2016).

These goods require frequent wire-based charging due to the restricted capacity of the batteries, which is inconvenient. Furthermore, patients must endure pain and expense when replacing batteries through surgery in order for implantable medical devices (IMDs) to function for an extended period of time. Considering all of these problems, creating a new power transfer method that eliminates cables is important. The use of WPT technology is the most likely fix. The transfer of electric energy without the use of wires is known as WPT, or conduitless energy transmission (Zeng et al., 2017).

Currently, the prominent concerns in the community revolve around the energy crisis and environmental pollution. Utilizing solar energy is an effective method to address these issues. Space-Based Solar Power (SBSP) has the benefit of higher conversion efficiency and immunity to meteorological conditions, unlike terrestrial solar installations. Solar-based power (SBSP) is an optimal solution for addressing both the environment pollution and energy crisis (Ma & Kb, 2021).

The transfer of electrical power from space to the Earth is a significant concern that requires attention. Microwave Power Transfer (MPT) can be a significant factor in this field. Electric Vehicles (EVs) have the potential to reduce the environmental concerns associated with traditional fuel vehicles. One of the primary concerns with electric vehicles is the fear of running out of power and the hassle of finding charging stations.

Wireless charging technology has the capacity to greatly improve the driving range and enhance the whole charging process for electric vehicles. Therefore, Wireless Power Transfer (WPT) technology plays an important role in tackling the issues related to shortages of energy and environmental damage (Aydin et al., 2022).

Wireless communication, or wireless signal transfer, releases people from the limitations of distance and place. It has significantly enhanced human welfare by greatly accelerating the advancement of associated technologies and promoting cross-cultural engagement. Similar to its energy-related cousin, WPT has the power to liberate people from wires and expand the use of electric energy, significantly altering people's lives (Kuresan et al., 2019).

Wireless power transfer (WPT) has a wide range of applications in various industries and sectors. It offers a convenient, cost-effective, and safe alternative to traditional wired power transfer methods. WPT can be used in challenging environments such as underwater or mines, implantable medical devices, rail trains, consumer electronics, electric vehicles (EVs), and high-voltage applications. Based on research conducted by Navigant, there is a projected increase in global revenue for wireless power from $1.3 billion in 2015 to $17.9 billion in 2024. Wireless power transfer (WPT) holds tremendous potential (Huda et al., 2022).

The World Economic Forum recognized WPT as one of the top 10 emerging technologies in 2012 and 2013 due to its wide range of applications and positive economic and social effects. In addition, the Massachusetts Institute of Technology (MIT) Technology Review named it one of the 10 breakthrough innovations of 2016 (Strassner & Chang, 2013).

Nikola Tesla conducted a number of experiments in the 19th century that are the origins of WPT. He demonstrated his wirelessly powered lightbulbs in 1893, which was a pioneering experiment in wireless power transfer. In 1900,

Nikola Tesla constructed a Wardenclyffe Tower, which is often referred to as the Tesla Tower. He intended to transmit power without the use of physical wires by enveloping the entire surface of the globe with electromagnetic waves of low frequency. Despite the failure of this effort, his concepts on wireless power transfer (WPT), including electromagnetic induction, tuning, and resonant circuits, have consistently served as a source of inspiration. These are the fundamental principles of modern magnetically coupled resonant wireless power transfer (WPT) technology (Hui, 2018).

Electrical energy was successfully transferred between two identical 100-MHz dipoles utilized for both transmitting and receiving by Westinghouse Laboratory researcher H. V. Noble in the 1930s. These dipoles were positioned around 8 meters apart and operated at power levels in the range of hundreds of watts. Peter Glaser established the idea of Space-Based Solar Power (SBSP) in 1968. Subsequently, the research on SBSP has been advanced (Wang et al., 2022).

1.1.3. Categories and Principles of Wireless Power Transfer (WPT)

WPT can be categorized according to the differences in principle, as seen in Figure 1.1. Magnetic induction, electromagnetic radiation, and electric induction. Electromagnetic wave/field phenomena are the foundation of WPT technologies. There are two distinct regions when it comes to the electromagnetic field generated by electromagnetic waves: the far field, also known as the radiation field, and the close field, further known as the induction field. The category is established by the distance among the source of the field and the point of observation, typically with one wavelength serving as the crucial threshold. The region beyond one wavelength is referred to as the far field, whereas the region within a distance much smaller than one wavelength is known as the near field (Erel et al., 2021).

WPT technology utilizing EM radiation operates on a far field. The transmitting equipment's actual length is significantly shorter than the transmission distance. Microwave WPT (also known as MPT) and laser WPT are the two technologies used in electromagnetic radiation WPT. Utilizing orienting devices can effectively guide the direction of electromagnetic waves, resulting in enhanced efficiency. Currently, due to its low efficiency and human hazard, it is in its early stages (Jiang et al., 2017).

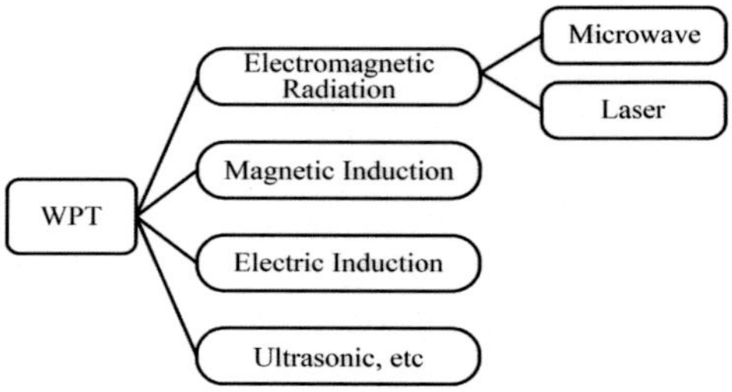

Figure 1.1. *Illustration of category of WPT.*

Source: Hui, SYR; Creative Commons License.

Magnetic induction and electric induction: The near field is the foundation of both WPT technologies. The transfer distance and physical length of the transmitting equipment are similar.

Through the process of magnetic induction WPT technology, often referred to as magnetically coupled WPT, gained considerable attention due to its exceptional transfer efficiency, extensive transfer range, and substantial transfer power capabilities. Magnetic induction between two linked coils is the foundation of magnetic induction WPT (Song et al., 2021).

Electromagnetic induction Wireless Power Transfer (WPT) enables the wireless transmission of electricity by utilizing the electric field generated by the metal plates on both the transmitter and receiver sides. Because of its low transfer distance and electric field hazard, it now receives far less attention than magnetic induction WPT technology.

There are various types of wireless power transfer (WPT) technologies available. These include ultrasonic WPT, which utilizes ultrasonic waves, as well as WPT that utilizes the hysteresis expansion or piezoelectric effect.

The fundamental concepts of magnetic induction wireless power transfer (WPT) are demonstrated as follows. The transmission coil generates an oscillating electric current that creates an oscillating magnetic field, some of which is connected to the reception coil. Power is delivered to the load when it is connected to the receiving coil. Figure 1.2 shows the concept of magnetic induction in wireless power transfer (WPT) (El Rayes et al., 2016).

There are different classifications of magnetic induction wireless power transfer (WPT). One classification distinguishes between resonant WPT and inductive WPT. There are differences between them in terms of the components used. Stray capacitors play a crucial role in resonant wireless power transfer

(WPT) by effectively resonating with the coils at a high frequency. On the other hand, in inductive WPT, lumped capacitors are used instead. However, the fundamental principles remain unchanged (Clerckx & Bayguzina, 2016).

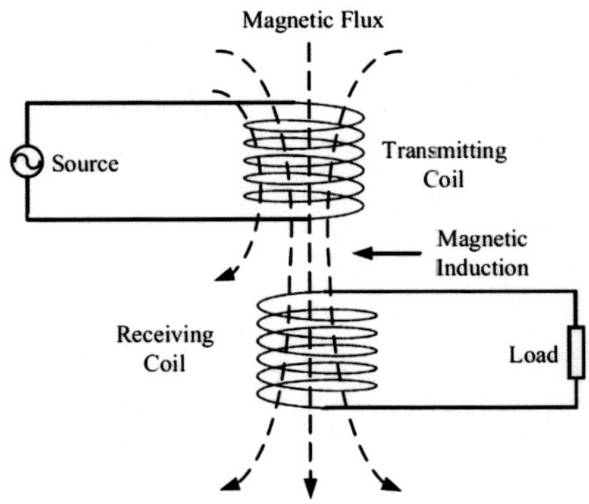

Figure 1.2. *Illustration of principle of magnetic induction WPT technology.*

Source: Zhong, W.; Creative Commons License.

1.2. BASIC STRUCTURE AND THEORY OF WIRELESS POWER TRANSFER

Learning Objectives:

- Understand the wireless power structure and theories.

1.2.1. WPT Structures

There are three primary methods for categorizing the structures of wireless power transfer (WPT). These methods include considering different resonant types, the presence of impedance matching coils, and the number of resonant coils (Yang et al., 2019).

1. **Two-Coil Structure and Four-Coil Structure:** As shown in Figure 1.3, WPT can be categorized into two-coil and four-coil configurations based on the presence of impedance matching coils. The four-coil configuration includes a source coil and a load coil, which sets it apart from the two-coil configuration. The load coil and source coil typically comprise smaller coils with a reduced number of turns (typically 1 or 2) or a smaller physical dimension. Both the load coil and source coil can isolate the load and source from the

receiving and sending coils, as well as adjust the impedance (Seo, 2019).

2. **SS, SP, S-LCL, PS, PP, P-LCL, LCL-S, LCL-P, and LCL-LCL:** Series (S), parallel (P), and inductor-capacitor-inductor (LCL) are the three types of resonant circuits. WPT can be categorized into nine structures, as shown in Figure 1.4, depending on the combination of various resonant types in receiver and transmitter: SS, SP, S-LCL, PS, PP, P-LCL, LCL-S, LCL-P, and LCL-LCL. The ideal load resistance, the receiving side's open-circuit and short-circuit characteristics, and the constant-current or load-independent constant-voltage characteristics are among the transfer characteristics that vary depending on the type of resonance (Namiki et al., 2023).

3. **Multiple-Transmitter, Intermediate-Coil Structure, and Receiver Structure:** WPT can be categorized into three types based on the number of resonators serving distinct purposes: multiple transmitters, multiple receivers, and numerous intermediate coils. As seen in Figure 1.5, power is transferred from n sources to a single load in the multiple-transmitter arrangement. High efficiency and a consistent magnetic field may be created over a large area with the multiple-transmitter configuration. Furthermore, the multiple-transmitter system's power output can be increased by precisely adjusting the phases of the various sources. As seen in Figure 1.6, power is delivered from a single source to n loads in the multiple-receiver system (Sharma et al., 2022). This structure is highly applicable in situations where there is a need for simultaneous charging of multiple loads, which is a frequent requirement in real-world scenarios. As shown in Figure 1.7, the multiple-intermediate-coil construction extends the transfer distance by inserting several relay resonators between the transmitter and the receiver. Moreover, the multiple-intermediate-coil structure can be used to realize a non-linear power transfer path. The multiple-transmitter multiple-receiver structure, the multiple-transmitter multiple-receiver structure and multiple-receiver structure using intermediate coils are examples of combinations of these three structures. They are not going to be presented here, as their structures are readily derived (Lee & Jang, 2017).

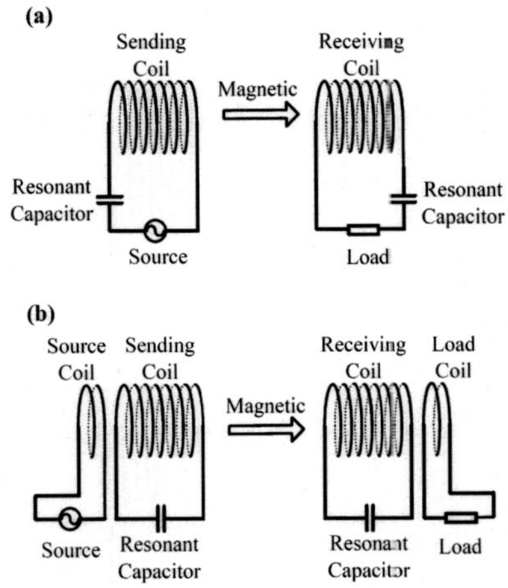

Figure 1.3. *Illustration of two basic structures: (a) two-coil structure; (b) four-coil structure.*

Source: Glaser, P.; Creative Commons License.

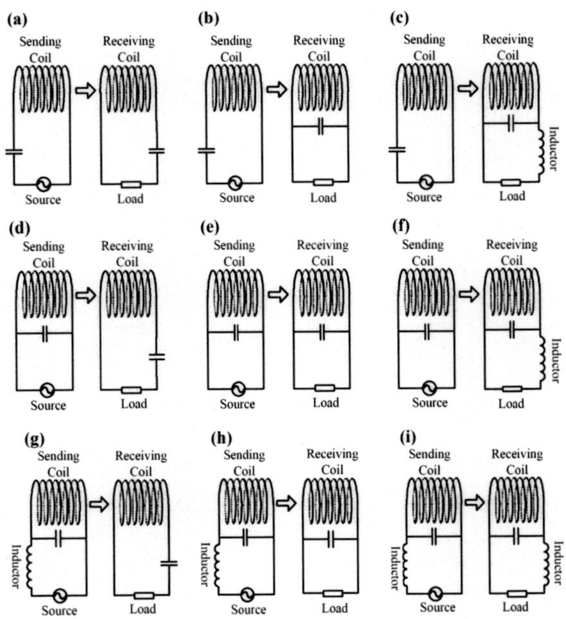

Figure 1.4. *Illustration of Nine basic structures: (a) SS; (b) SP; (c) S-LCL; (d) PS; (e) PP; (f) P-LCL; (g) LCL-S; (h) LCL-P; and (i) LCL-LCL.*

Source: Huang, C.; Creative Commons License.

1.2.2. WPT Theories

A WPT system can be explained using three fundamental theories: Band Pass Filter (BPF), Coupled Mode Theory (CMT), and Circuit Theory (CT) (Clerckx et al., 2022).

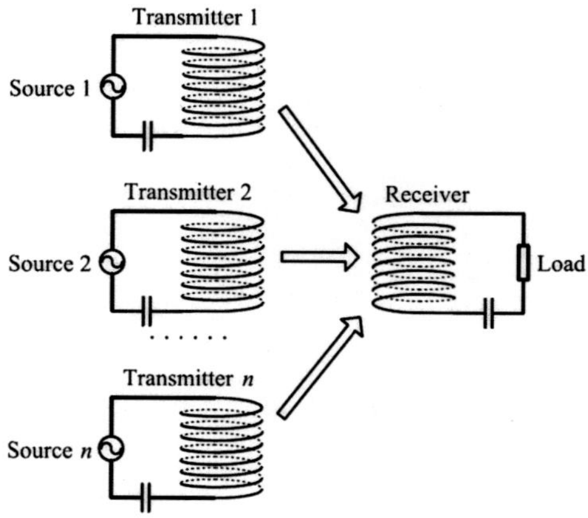

Figure 1.5. *Illustration of multiple-transmitter structure.*

Source: Covic, G. A.; Creative Commons License.

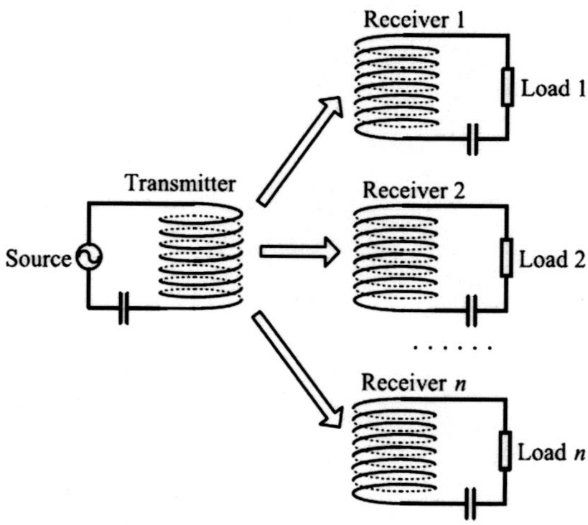

Figure 1.6. *Multiple-receiver structure.*

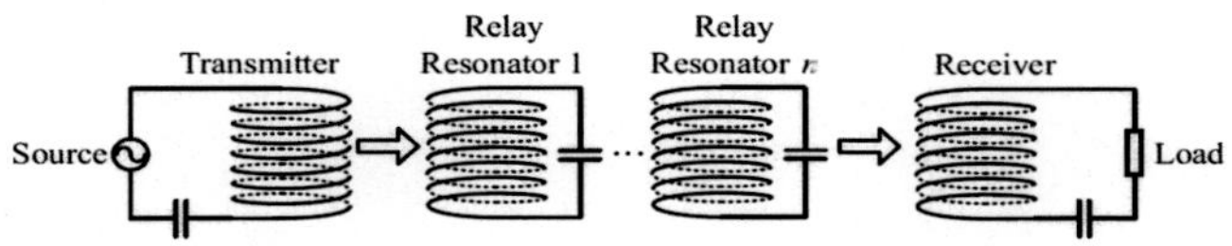

Figure 1.7. *Multiple-intermediate-coil structure.*

Source: Green, A. W.; Creative Commons License.

1. **CMT:** The CMT technique is commonly used to describe the vibrational systems coupling in time or space. It proves to be particularly valuable in the analysis of coupled resonators in the field of wireless power transfer (Selvi et al., 2018). Figure 1.3 shows CMT model that can be defined as

$$\begin{cases} \frac{da_1}{d_t} = (j\omega_0 - \Gamma_1)\boldsymbol{a}_1 + j\kappa_{12}\boldsymbol{a}_2 + A_S e^{j\omega_0 t} \\ \frac{da_2}{d_t} = [j\omega_0 - (\Gamma_2 + \Gamma_L)]\boldsymbol{a}_2 + j\kappa_{21}\boldsymbol{a}_1 \end{cases} \quad (1.1)$$

Throughout the thesis, the matching phasors are represented by the parameters in bold.

Eqn. (1.1) can be rewritten as:

$$\frac{d}{dt}\begin{bmatrix} \boldsymbol{a}_1 \\ \boldsymbol{a}_2 \end{bmatrix} = \begin{bmatrix} j\omega_0 - \Gamma_1 & j\kappa_{12} \\ j\kappa_{21} & j\omega_0 - (\Gamma_2 + \Gamma_L) \end{bmatrix}\begin{bmatrix} \boldsymbol{a}_1 \\ \boldsymbol{a}_2 \end{bmatrix} + \begin{bmatrix} A_S e^{j\omega_0 t} \\ 0 \end{bmatrix} \quad (1.2)$$

By solving Eqn. (1.2), one can obtain the relevant parameters of the system.

2. **BPF:** The WPT system is considered to be a second-order band pass filter in the BPF. The WPT system is designed in the same manner as a second-order band pass filter. Usually, filters like Butterworth or Chebyshev are used (Yamazaki et al., 2015).
3. **CT:** It is widely used and appreciated for its intuitive and simple nature. By utilizing the principles of Kirchhoff Current Law (KCL) and Kirchhoff Voltage Law (KVL), it is possible to develop a comprehensive model for describing a wireless power transfer (WPT) system (Goldman, 2007) Consider Figure 1.3(a) as an example. The equivalent circuit is shown in Figure 1.8.

The relationship between L2, L3, and M23 is:

$$M_{23} = \rho_{23}\sqrt{L_2 L_3} \quad (1.3)$$

where; ρ_{23} satisfies 0 q23 1.

According to KVL, the model shown in Figure 1.8 can be constructed by:

$$\begin{cases} U_S = \left(R_2 + j\omega L_2 + \frac{1}{j\omega C_2}\right) I_2 + j\omega M_{23} I_3 \\ 0 = j\omega M_{23} I_2 + \left(R_3 + j\omega L_3 + \frac{1}{j\omega C_3} + R_L\right) I_3 \end{cases} \quad (1.4)$$

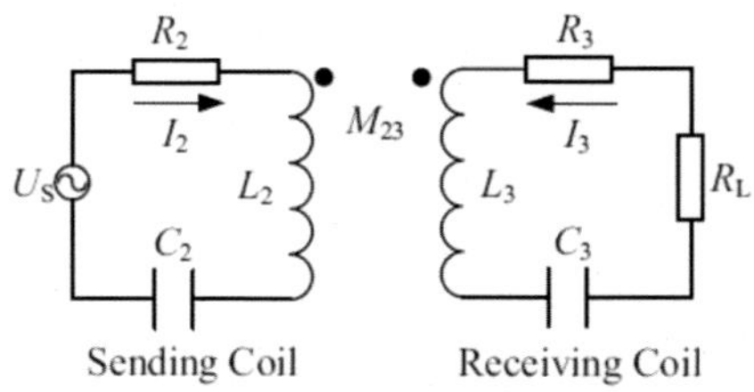

Figure 1.8. *Illustration of equivalent circuit of two-coil structure with controlled sources.*

Source: Green, A. W.; Creative Commons License.

By separating the interconnected inductances in Figure 1.8, it is possible to derive many equivalent circuits. Figure 1.9 shows the equivalent circuit with controlled sources. The receiving coil current I_3 determines the regulated voltage source on the sending coil, $j\omega M_{23} I_3$. In the same way, the sending coil current I_2 determines the regulated voltage source on the receiving coil, which is j $j\omega M_{23} I_2$. This circuit is not appropriate for solving the model (Maglinte et al., 2007).

$$\begin{cases} U_S = \left(R_2 + j\omega (L_2 - M_{23}) + \frac{1}{j\omega C_2}\right) I_2 + j\omega M_{23}(I_2 + I_3) \\ 0 = j\omega M_{23}(I_2 + I_3) + \left(R_3 + j\omega (L_3 - M_{23}) + \frac{1}{j\omega C_3} + R_L\right) I_3 \end{cases} \quad (1.5)$$

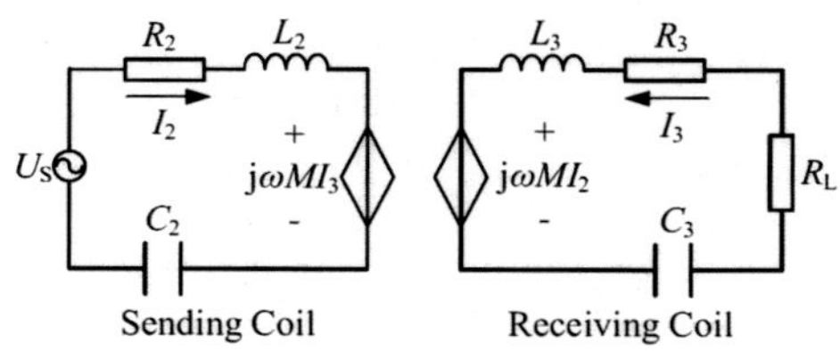

Figure 1.9. *Representation of equivalent circuit of two-coil structure with controlled sources.*

Source: Karalis, A.; Creative Commons License.

Eqn. (1.5) can be used to calculate the T-type equivalent circuit for the two-coil design, which is shown in Figure 1.10. The solution to this equivalent circuit can be derived by using the principles of series and parallel connections of impedances (Townsend et al., 2004).

It should be noted that $L_2 - M_{23}$ and $L_3 - M_{23}$ do not represent the leakage inductances. When the value of L_2 –> L_3 and the value of $\rho_{23} > \sqrt{L_3/L_2}$ multiplied by $L_2 - M_{23} < 0$, the result is less than zero. Comparable outcomes can be achieved when $L_3 - M_{23} < 0$. As seen in Figure 1.11, a T-type equivalent circuit with a turn ratio has been created to solve the problem of negative inductance values and effectively represent the leakage inductances (Stamm & Nagel, 2002).

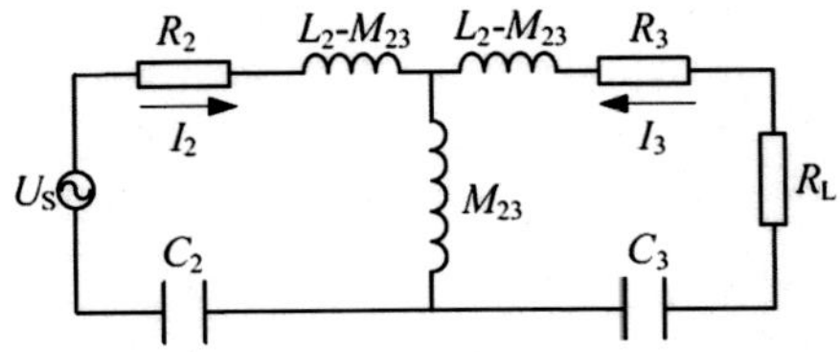

Figure 1.10. *Illustration of T-type equivalent circuit of two-coil structure.*

Source: Soljacic, M.; Creative Commons License.

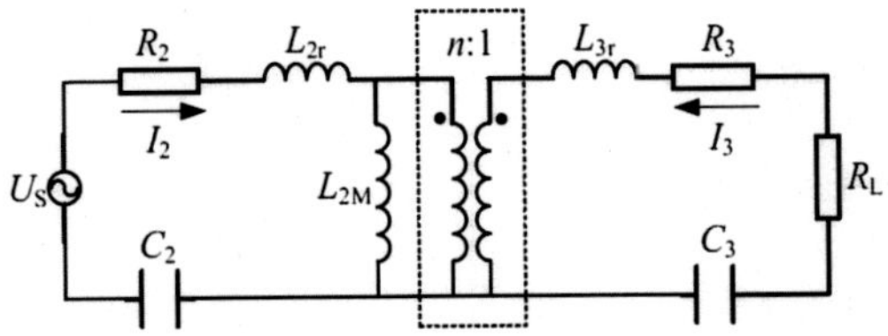

Figure 1.11. *Illustration of T-type equivalent circuit with a turn ratio of two-coil structure.*

Source: Moffatt, R.; Creative Commons License.

Eliminating I_3 in Eqn. (1.4) yields:

$$\begin{aligned} U_S &= \left(R_2 + j\omega L_2 + \frac{1}{j\omega C_2}\right)\boldsymbol{I}_2 + \frac{(\omega M_{23})^2}{R_3 + j\omega L_3 + \frac{1}{j\omega C_3} + R_L}\boldsymbol{I}_2 \\ &= \left(R_2 + j\omega L_2 + \frac{1}{j\omega C_2}\right)\boldsymbol{I}_2 + Z_{ref23}\boldsymbol{I}_2 \end{aligned} \quad (1.6)$$

where; Z_{ref23}, the reflected impedance between the sending and receiving coils, is defined as:

$$Z_{ref23} = \frac{(\omega M_{23})^2}{R_3 + j\omega L_3 + \frac{1}{j\omega C_3} + R_L} \tag{1.7}$$

By examining Eqn. (1.6), The receiving loop has a noticeable impact as it introduces a varying impedance that is linked in series with the sending coil branch. The calculation of the variable impedance, as demonstrated in Eqn. (1.7), requires the squaring of the mutual inductance and its subsequent inverse relationship with the receiving coil and loop impedance. Figure 1.12 displays the circuit diagram that includes the reflected impedance of the two-coil arrangement. The reflected impedance and the controlled voltage on the receiving side can be used to calculate the transmitter current, which generates the receiving coil current (Mettler et al., 2000) (Figure 1.13).

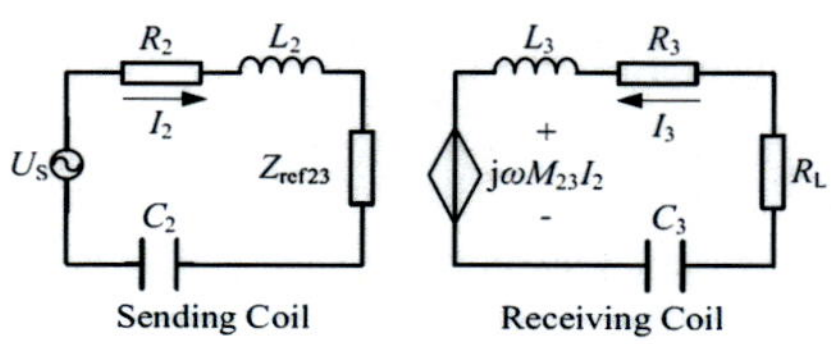

Figure 1.12. *Illustration of equivalent circuit with reflected impedance o two-coil structure.*

Source: Soljacic, M.; Creative Commons License.

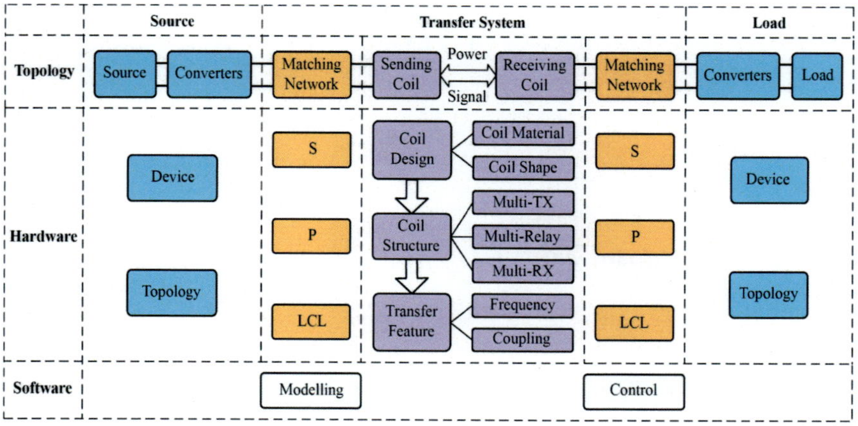

Figure 1.13. *Illustration of research focus and components of WPT system.*

Source: Lee, C. K.; Creative Commons License.

1.3. METHODS OF WIRELESS POWER TRANSFER

Learning Objectives:

- Understand various methods of wireless power transfer, including inductive and resonance coupling.

Wireless power transmission can be accomplished through a variety of approaches, including designs for both near-field and far-field wireless power transfer (Mahesh et al., 2021). These techniques are covered in subsections.

1.3.1. Inductive Coupling

Non-radiative near-field energy transfer is possible with inductive coupling as a wireless power transfer technique. The technique uses electromagnetic induction concepts, much like a transformer. Here, the primary and secondary coils are used in the design. The secondary coil induces a current by means of a time-varying magnetic field produced by the first coil. The fact that inductive coupling eliminates the need for connections between the power transmitter and power receiver is one benefit of using it for wireless power transfer (Miura et al., 2005).

Practice Problem

In an inductive coupling WPT system, the primary coil has 200 turns and an inductance of 5 mH. If the secondary coil has 100 turns, what will be its induced voltage when the primary coil is supplied with 10 V?

Solutions to Practice Problem

Primary coil turns (N1) = 200
Primary coil inductance (L1) = 5 mH = 5×10^{-3} H
Secondary coil turns (N2) = 100
Supply voltage (V1) = 10 V
Using the formula for induced voltage in a transformer: V2 = (N2/N1) × V1,
V2 = (100/200) × 10
V2 = 5 V.

1.3.2. Resonance Coupling

Non-radiative near-field energy transfer is possible with inductive coupling as a wireless power transfer technique. The technique uses electromagnetic induction concepts, much like a transformer. Here, the primary and secondary coils are used in the design. The secondary coil induces a current by means of a time-varying magnetic field produced by the first coil. The fact that inductive coupling eliminates the need for connections between the power transmitter and power receiver is one benefit of using it for wireless power transfer (Beh et al., 2012).

1.3.3. Power Transfer by Radio Waves and Optical Waves

Microwaves and radio waves can be used for far-field wireless power transfer. These waves transfer power radiatively and have high frequencies and small wavelengths. In electric circuits, a rectenna is essentially constructed using a receiver circuit that consists of an antenna and a rectifier circuit.

In order to convert alternating electromagnetic power to DC power, the rectifier circuit is helpful. Using radio waves, the rectenna facilitates wireless power transfer. The ability of radio waves or microwaves to travel at speeds up to the speed of light makes them useful for wireless power transfer. This allows power to be transmitted over long distances (Bagci et al., 2014).

Utilizing stimulated emission of radiation (also known as laser amplification), light can be amplified in order to transfer power via optical waves. Using this technique, power is sent by turning energy into a laser beam that is pointed in the direction of a solar cell receiver (Pinheiro et al., 2000).

1.4. CHARGING AND WPT SYSTEMS IN ELECTRIC VEHICLES

Learning Objectives:

- Understand the three categories of WPT systems for electric vehicles.

The wireless charging systems used for WPT can be categorized into three groups based on the mobility status of the electric vehicle (EV),

- Stationary WPT (SWPT);
- Semi or quasi dynamic WPT (QDWPT);
- Dynamic WPT (DWPT).

The Stationary Wireless Power Transfer (SWPT) system enables the charging of an Electric Vehicle (EV) by utilizing the Wireless Power Transfer (WPT) technology. The SWPT differs from the plug-in wired charger solely in its wireless energy transfer capability. The SWPT system is highly effective in urban environments, particularly for the installation of stations dedicated to electric vehicle (EV) charging. The SWPT technology utilizes a pair of pads.

One component is designed for reception and is put on the vehicle, while the other component is intended for transmission and is installed on the charger (Machura & Li, 2019).

The QDWPT, or Quick Dynamic Wireless Power Transfer, is specifically designed for rapid charging over a short period of time. This is why it is commonly put at bus and taxi stations. This system can utilize temporary stops at traffic lights to recharge electric vehicles (EVs). In a Dynamic Wireless Power Transfer (DWPT) system, electrical energy is transmitted to an Electric Vehicle (EV) wirelessly while the vehicle is moving. There is no need to stop and stay in one location for the automobile to recharge.

The QDWPT, DWPT, and SWPT can be achieved using either the IPT or the CPT. Currently, commercial scale deployments prefer using the IPT (Inductive Power Transfer) since it has a greater power transfer capability and can work over bigger distances between the transmitter and receiver compared to the CPT (Capacitive Power Transfer). Nevertheless, the implementation of the IPT for the SWPT, QDWPT, and DWPT requires the use of sizable copper coils with ferrous cores, resulting in a system which is heavy (Bagchi et al., 2021).

In contrast, prototypes and implementations based on CPT utilize fewer complex structures that consist of plates and foils. Not only is the resulting charging mechanism lighter, but it also uses less energy. However, the power transfer levels are diminished and the distance between the objects is rather minimal. Ongoing research aims to address the limitations of the QDWPT, CPT-based SWPT, and DWPT charging systems, as they are considered simple and cost-effective (Figure 1.14).

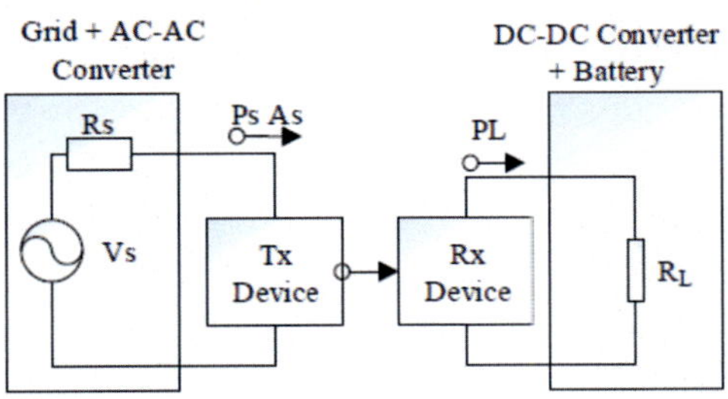

Figure 1.14. *Illustration of a wireless charging system for an electric vehicle.*

The charging procedure of electric vehicle (EV) batteries consists of two stages. One form of operation involves maintaining a steady charging current while allowing the charging voltage to fluctuate. The charge is commonly known as the constant current (CC) charge. In the alternate mode, the charging voltage remains constant, but the charging current is permitted to vary. This is known as the constant voltage (CV) charge. During the initial stage of recharging a fully depleted battery, the Constant Current (CC) mode is employed. The charging current is typically set at a percentage of the battery's rated capacity (often between 10% to 30% of the battery's total capacity), depending on the battery chemistry and the manufacturer's recommendations. Once the battery State of Charge (SoC) approaches its rated capacity, typically about 80%, the charging process switches to Constant Voltage (CV) mode.

This guarantees a seamless charging procedure without the risk of overheating and causing harm to the battery; however it may necessitate a longer duration for charging. Using the CV mode to charge a fully drained battery initially can lead to a high charging current, potentially causing overheating or damage to the battery. Hence, the CV mode is typically avoided while charging a battery to a level below 80% of its total capacity. Similarly, utilizing the CC mode to charge a battery close to its rated capacity might lead to overcharging, potentially causing overheating or damage to the battery. The variability of the electrical current and voltage diminishes the longevity of the battery. If there are changes in the load throughout the battery charging process, which leads to a change in its State of Charge (SoC), the Constant Current (CC) and Constant Voltage (CV) modes can be employed as necessary to ensure an efficient charging process (Razu et al., 2021).

1.5. APPLICATION OF WIRELESS POWER TRANSFER

Learning Objectives:

- Understand the potential applications and impact of wireless power transfer.

The potential of wireless power transfer to transform various industries is truly remarkable. From the simplicity of charging a cell phone to the potential of powering the entire Earth, the possibilities of wireless charging are vast. Consumers can expect to see the initial applications of a charging station with a range of approximately one to five meters. This compact device has the capability to wirelessly charge electronic devices that are compatible with its system, providing convenient and efficient charging. Wireless power transfer charging systems have been extensively studied and have demonstrated efficiencies comparable to traditional charging devices. As an example, a typical household would require one transmitter for each room in order to achieve a fully wireless environment (Basar et al., 2014).

This will just provide convenience to consumers and will not offer any additional services. Electronic corporations may be resistant to this new approach as it would prevent them from producing a unique charger for each of their products and compel users to purchase multiple charges. An inherent benefit of personal gadgets that integrate wireless chargers is their ability to self-charge whenever they enter the vicinity of a power transmitter, eliminating the need for manual charging. As wireless power transmission technology advances, the range of potential applications can expand. The extent of the electromagnetic waves will begin to expand (Pugi et al., 2019).

When the distance reaches approximately 20 meters, a single transmitter positioned near a home will have the capability to charge whole buildings. As the coverage area increases, each transmitter will have the capability to supply electricity to entire blocks and streets. Once this milestone is reached, consumers will observe a transition in cars to become fully electric, equipped with the ability to be charged wirelessly. Infrastructure will be modified to incorporate strategically spaced wireless chargers on roads, enabling vehicles to operate continuously without running out of power.

The last stage of this phase involves the utilization of solar panels in outer space. Solar powered satellites will utilize outer space solar panels to gather the Sun's energy and transmit it to Earth. This will significantly diminish the dependence on traditional fuels. This will provide a limitless supply of energy. Currently, in the field of wireless power transfer, consumers will observe fluctuations in the stock market related to commodities such as petroleum, natural gas, and coal. Residences and automobiles will become independent from having the need for heating or energy derived from these sources (Rao & Chiao, 2015).

Global warming and the world's reliance on oil are the two main issues. Both issues will be treated using this system. Even though it will take many years to finish, everyone should start considering this as a long-term fix. The energy crisis will be resolved by wireless power transfer (Samanta & Rathore, 2016).

SUMMARY

- The chapter offers a thorough and detailed overview of the field. The chapter begins by providing an overview of Wireless Power Transfer (WPT) and its historical background. The chapter provides an overview of various categories and principles of Wireless Power transmission (WPT), such as inductive coupling, resonance coupling, and power transmission by radio waves and optical waves.
- It discusses numerous WPT structures and theories while examining the fundamental structure and theories underpinning WPT systems. The chapter also explores wireless power transfer techniques and their uses, with particular attention to areas like electric car charging.
- Additionally, it discusses the operational concepts and charging procedure associated with WPT systems, concluding in an analysis of practical implementations of WPT technology.

REVIEW QUESTIONS

1. Explain inductive and resonance coupling principles in WPT, contrasting them.
2. What are the categories of WPT systems and their applications?
3. Describe the basic structure and theory of a WPT system.
4. Discuss methods of wireless power transfer and their advantages.
5. Explore WPT applications, particularly in electric vehicle charging.
6. Explain WPT operating principles during charging and factors affecting efficiency.
7. Reflect on current advancements and future prospects in WPT.

MULTIPLE CHOICE QUESTIONS

1. What is one of the fundamental principles of Wireless Power Transfer (WPT)?

a. Ohm’s Law

b. Faraday’s Law

c. Newton’s Law

d. Boyle’s Law

2. Which of the following is NOT a category of WPT?

a. Inductive Coupling

b. Mechanical Coupling

c. Resonance Coupling

d. Power Transfer by Radio Waves

3. Which method of WPT utilizes electromagnetic fields to transfer power?

a. Inductive Coupling

b. Resonance Coupling

c. Power Transfer by Radio Waves

d. Optical Waves

4. What is a common application of WPT discussed in the chapter?

a. Wireless data transfer

b. Wireless audio transmission

c. Charging in electric vehicles

d. Satellite communication

5. What is the basic structure of a WPT system?

a. Transmitter and receiver

b. Capacitor and resistor

c. Battery and generator

d. Amplifier and antenna

Answers to Multiple Questions

1. (b); 2. (b); 3. (a); 4. (c); 5. (a)

REFERENCES

1. Aydin, E., Aydemir, M. T., Aksoz, A., El Baghdadi, M., & Hegazy, O. (2022). Inductive power transfer for electric vehicle charging applications: A comprehensive review. *Energies, 15*(14), 4962.
2. Bagchi, A. C., Kamineni, A., Zane, R. A., & Carlson, R. (2021). Review and comparative analysis of topologies and control methods in dynamic wireless charging of electric vehicles. *IEEE Journal of Emerging and Selected Topics in Power Electronics, 9*(4), 4947–4962.
3. Bagci, T., Simonsen, A., Schmid, S., Villanueva, L. G., Zeuthen, E., Appel, J., & Polzik, E. S. (2014). Optical detection of radio waves through a nanomechanical transducer. *Nature, 507*(7490), 81–85.

4. Basar, M. R., Ahmad, M. Y., Cho, J., & Ibrahim, F. (2014). Application of wireless power transmission systems in wireless capsule endoscopy: An overview. *Sensors, 14*(6), 10929–10951.

5. Beh, T. C., Kato, M., Imura, T., Oh, S., & Hori, Y. (2012). Automated impedance matching system for robust wireless power transfer via magnetic resonance coupling. *IEEE Transactions on Industrial Electronics, 60*(9), 3689–3698.

6. Clerckx, B., & Bayguzina, E. (2016). Waveform design for wireless power transfer. *IEEE Transactions on Signal Processing, 64*(23), 6313–6328.

7. Clerckx, B., Kim, J., Choi, K. W., & Kim, D. I. (2022). Foundations of wireless information and power transfer: Theory, prototypes, and experiments. *Proceedings of the IEEE, 110*(1), 8–30.

8. El Rayes, M. M., Nagib, G., & Abdelaal, W. A. (2016). A review on wireless power transfer. *International Journal of Engineering Trends and Technology (IJETT), 40*(5), 272–280.

9. Erel, M. Z., Bayindir, K. C., Aydemir, M. T., Chaudhary, S. K., & Guerrero, J. M. (2021). A comprehensive review on wireless capacitive power transfer technology: Fundamentals and applications. *IEEE Access, 10*, 3116–3143.

10. Goldman, L. W. (2007). Principles of CT and CT technology. *Journal of Nuclear Medicine Technology, 35*(3), 115–128.

11. Huda, S. A., Arafat, M. Y., & Moh, S. (2022). Wireless power transfer in wirelessly powered sensor networks: A review of recent progress. *Sensors, 22*(8), 2952.

12. Hui, S. R. (2016). Past, present and future trends of non-radiative wireless power transfer. *CPSS Transactions on Power Electronics and Applications, 1*(1), 83–91.

13. Hui, S. R. (2018). Technical and safety challenges in emerging trends of near-field wireless power transfer industrial guidelines. *IEEE Electromagnetic Compatibility Magazine, 7*(1), 78–86.

14. Jiang, C., Chau, K. T., Liu, C., & Lee, C. H. (2017). An overview of resonant circuits for wireless power transfer. *Energies, 10*(7), 894.

15. Kuresan, H., Samiappan, D., & Masunda, S. (2019). Fusion of WPT and MFCC feature extraction in Parkinson's disease diagnosis. *Technology and Health Care, 27*(4), 363–372.

16. Lee, S. B., & Jang, I. G. (2017). Layout optimization of the receiver coils for Multi transmitter wireless power transfer systems. *IEEE Journal of Emerging and Selected Topics in Power Electronics, 5*(3), 1311–1321.

17. Li, S., & Mi, C. C. (2014). Wireless power transfer for electric vehicle applications. *IEEE Journal of Emerging and Selected Topics in Power Electronics, 3*(1), 4–17.
18. Ma, D., & Kb, R. (2021). Systematic literature review on wireless power transmission. *Turkish Journal of Computer and Mathematics Education, 12*, 4400–4406.
19. Machura, P., & Li, Q. (2019). A critical review on wireless charging for electric vehicles. *Renewable and Sustainable Energy Reviews, 104*, 209–234.
20. Maglinte, D. D., Sandrasegaran, K., Lappas, J. C., & Chiorean, M. (2007). CT enteroclysis. *Radiology, 245*(3), 661–671.
21. Mahesh, A., Chokkalingam, B., & Mihet-Popa, L. (2021). Inductive wireless power transfer charging for electric vehicles–A review. *IEEE Access, 9*, 137667–137713.
22. Mettler, F. A., Wiest, P. W., Locken, J. A., & Kelsey, C. A. (2000). CT scanning: Patterns of use and dose. *Journal of Radiological Protection, 20*(4), 353.
23. Miura, N., Mizoguchi, D., Sakurai, T., & Kuroda, T. (2005). Analysis and design of inductive coupling and transceiver circuit for inductive inter-chip wireless super connect. *IEEE Journal of Solid-State Circuits, 40*(4), 829–837.
24. Namiki, H., Imura, T., & Hori, Y. (2023). Comparison of multiple circuits including LCL in inductive power transfer and capacitive power transfer. In *IECON 2023-49th Annual Conference of the IEEE Industrial Electronics Society* (Vol. 2, pp. 1–7).
25. Patil, D., McDonough, M. K., Miller, J. M., Fahimi, B., & Balsara, P. T. (2017). Wireless power transfer for vehicular applications: Overview and challenges. *IEEE Transactions on Transportation Electrification, 4*(1), 3–37.
26. Pinheiro, F. A., Martinez, A. S., & Sampaio, L. C. (2000). Vanishing of energy transport velocity and diffusion constant of electromagnetic waves in disordered magnetic media. *Physical Review Letters, 85*(26), 5563.
27. Pugi, L., Reatti, A., & Corti, F. (2019). Application of modal analysis methods to the design of wireless power transfer systems. *Meccanica, 54*, 321–331.
28. Rao, S., & Chiao, J. C. (2015). Body electric: Wireless power transfer for implant applications. *IEEE Microwave Magazine, 16*(2), 54–64.
29. Razu, M. R. R., Mahmud, S., Uddin, M. J., Islam, S. S., Bais, B., Misran,

N., & Islam, M. T. (2021). Wireless charging of electric vehicle while driving. *IEEE Access, 9*, 157973–157983.

30. Samanta, S., & Rathore, A. K. (2016). Wireless power transfer technology using full-bridge current-fed topology for medium power applications. *IET Power Electronics, 9*(9), 1903–1913.
31. Selvi, S., Vishvaksenan, A., & Rajasekar, E. (2018). Cold metal transfer (CMT) technology-An overview. *Defence Technology, 14*(1), 28–44.
32. Seo, D. W. (2019). Comparative analysis of two-and three-coil WPT systems based on transmission efficiency. *IEEE Access, 7*, 151962–151970.
33. Sharma, S. K. P. R., Kar, D. P., & Bhuyan, S. (2022). Systematic literature review on wireless power transmission to multiple devices. *NeuroQuantology, 20*(9), 3082.
34. Song, M., Jayathurathnage, P., Zanganeh, E., Krasikova, M., Smirnov, P., Belov, P., & Krasnok, A. (2021). Wireless power transfer based on novel physical concepts. *Nature Electronics, 4*(10), 707–716.
35. Stamm, G., & Nagel, H. D. (2002). CT-expo–A novel program for dose evaluation in CT. *RoFo: Fortschritte auf dem Gebiete der Rontgenstrahlen und der Nuklearmedizin, 174*(12), 1570–1576.
36. Strassner, B., & Chang, K. (2013). Microwave power transmission: Historical milestones and system components. *Proceedings of the IEEE, 101*(6), 1379–1396.
37. Townsend, D. W., Carney, J. P., Yap, J. T., & Hall, N. C. (2004). PET/CT today and tomorrow. *Journal of Nuclear Medicine, 45*(1 suppl), 4S–14S.
38. Wang, D. A., Zhang, J., Cui, S., Bie, Z., Song, K., Zhu, C., & Matveevich, M. I. (2022). Modern advances in magnetic materials of wireless power transfer systems: A review and new perspectives. *Nanomaterials, 12*(20), 3662.
39. Yamazaki, E., Yamashita, N., Taniyasu, S., Lam, J., Lam, P. K., Moon, H. B., & Kannan, K. (2015). Bisphenol A and other bisphenol analogues including BPS and BPF in surface water samples from Japan, China, Korea and India. *Ecotoxicology and Environmental Safety, 122*, 565–572.
40. Yang, L., Li, X., Liu, S., Xu, Z., Cai, C., & Guo, P. (2019). Analysis and design of three-coil structure WPT system with constant output current and voltage for battery charging applications. *IEEE Access, 7*, 87334–87344.
41. Zeng, Y., Clerckx, B., & Zhang, R. (2017). Communications and signals design for wireless power transmission. *IEEE Transactions on Communications, 65*(5), 2264–2290.
42. Zhang, Z., Pang, H., Georgiadis, A., & Cecati, C. (2018). Wireless power transfer—An overview. *IEEE Transactions on Industrial Electronics, 66*(2), 1044–1058.

CHAPTER 2

FUNDAMENTALS OF WIRELESS POWER TRANSFER

LEARNING OBJECTIVES

At the end of this chapter, readers will be able to:

1. Understand the principles and classifications of Wireless Power Transfer (WPT) technologies.
2. Explore the key features and operational principles of inductive, magnetic resonance, capacitive, microwave, and optical WPT technologies.
3. Analyze the advantages and limitations of different WPT technologies in near field, mid-range, and far-field applications.
4. Identify key technologies and considerations in real commercial WPT systems.
5. Understand receiver localization methods and foreign object detection in WPT.
6. Analyze recent advances and future prospects in WPT technology.

INTRODUCTORY EXAMPLE

WiTricity Corporation, a leader in wireless power transfer technology, has been instrumental in advancing the field of inductive and magnetic resonance-based wireless power systems. Their technology enables efficient and convenient power transfer without the need for physical connectors. This is particularly evident in the wireless charging of electric vehicles (EVs), where WiTricity's solutions have simplified the charging process and improved user experience.

In a typical wireless charging setup for EVs, a charging pad is installed on the ground and connected to a power source. The EV, equipped with a receiver coil on its underside, is parked over the charging pad. When the system is activated, the charging pad generates an alternating magnetic field. This magnetic field induces an electric current in the receiver coil of the EV. The induced current is then converted into direct current (DC) to charge the vehicle's battery.

WiTricity's technology uses magnetic resonance to enhance the efficiency and range of this power transfer. By tuning both the transmitter and receiver coils to the same resonant frequency, the system achieves higher efficiency compared to traditional inductive methods. This advancement not only makes charging more efficient but also allows for greater flexibility in the positioning of the vehicle relative to the charging pad.

WiTricity's partnerships with major automakers have led to the development and integration of these wireless charging systems in various EV models. This collaboration demonstrates the practical application of wireless power transfer technology and its potential to transform the electric vehicle-charging infrastructure, making it more user-friendly and efficient.

UNIT INTRODUCTION

The technique known as wireless power transfer, or WPT, uses one or more transmitters to produce electromagnetic waves that are then processed by receivers with no kind of conductor to extract power from the waves. Unlike wireless communication systems, the receiver in WPT systems uses electromagnetic wave to power devices or store energy in a battery (Erel et al., 2021).

Nikola Tesla, an engineer, conducted the first wireless power transfer experiments towards the end of the 1800s. He succeeded in using radio waves to transfer power between two objects that were 48 kilometers apart. One of Tesla's previous tests involved using a 25-mile power supply to power 200 cordless lightbulbs. Safety concerns about electricity and humans were not taken into account for this research (Clerckx et al., 2018).

The research community did not become interested in WPT systems again until the 21st century (Patil et al., 2017). The rise in interest was driven by the emergence of power converters during that period, enabling the utilization of frequencies ranging from several kilohertz to kilowatt-level operations. The technology behind this new trend was first called Contactless Energy Transfer (CET). But in the end, "wireless power transfer" started to be used interchangeably (Zhang et al., 2018).

This chapter delves into the basics of Wireless Power Transfer (WPT), covering its different classifications, technologies, and applications. Let's start by exploring the various methods of WPT, which encompass inductive, capacitive, microwave, magnetic resonance, and optical technologies (Zeng et al., 2017). It is important to grasp the underlying principles behind each approach. In addition, one will explore the importance of related technologies like receiver localization, data communication, and foreign object detection in enhancing the efficiency and safety of WPT systems (Lu et al., 2015).

2.1. CLASSIFICATION CRITERIA FOR WIRELESS POWER TRANSFER (WPT) SYSTEMS

Learning Objectives:

- Understand the criteria for wireless power transfer systems based on criteria such as power, directionality, and medium.

WPT technology has become a tangible and practical solution. This technique is utilized in various commercial items, including mobile phones power mats, electric toothbrushes, and electric vehicles (EVs) charges (Detka & Górecki, 2022).

In 2017, products with this feature were shipped worldwide, mostly in smartphones, small home appliances and smart watches. This statistic indicated a 75% surge in sales compared to the previous year's recorded sales. This substantial surge is expected to last in the immediate future. This market sector is expected to grow to over 2.2 billion units by 2023, according to IHS Markit. Additionally, there will be significant economic benefits from the growth. According to Navigant Research, the revenue generated from wireless chargers is projected to reach approximately 17.9 billion dollars by 2024 (Jiang et al., 2017).

When examining WPT applications, they are implemented in various ways. As a result, systems for wireless power transfer can be categorized based on the following criteria (*see* Figure 2.1):

1. **Transferred Power:** Wireless power transfer (WPT) systems consist of various applications that involve transmitting different

power levels including Low power (up to 1 kW), medium power (between 1 and 100 kW), and the high power (beyond 100 kW). The power demand of the application has a significant impact on the design of the system (Raso et al., 1999). Efficiency is therefore less important in low-power applications than it is in other kinds of systems. Typically, the focus of low-power applications is to achieve the highest power transfer possible.

2. **Power Transfer Uni-Directional or Bi-Directional:** This criterion allows one to distinguish between WPT systems, where power transfer always begins at a fixed element that is coupled to a source. This idea shows a one-way wireless power transmission technique. As an alternative, there are bi-directional systems where the source receives energy from time to time from the load, such a battery or capacitor (Shevchenko et al., 2019).
3. **Gap:** The distance between the receiver and energy transmitter is referred to by this term. In certain applications, contact between these two components is necessary, even though all WPT systems avoid using wires between them. Power mats are one example of this. On the other hand, in certain applications, the distance between the receiver and the transmitter is few centimeters or even meters (Howell et al., 2001).
4. **Capacity to Function with Intermediate Objects:** This within the gap separating the receiver and power transmitter. Certain technologies cannot function properly when there are intermediate objects present, while others experience significant degradation in their performance. Nevertheless, there are also technologies in which the presence of these objects does not significantly affect the outcome (Mahesh et al., 2021).
5. **Number of Transmitters:** A basic configuration for a WPT system involves a single receiver as well as single power transmitter. To improve the spatial operability of WPT, it is possible to set up multiple transmitters in a specific area to transmit power to a load. Multiple transmitters can be activated at the same time, taking into account their efficiency of power transfer and power availability. For example, this can be done by utilizing renewable energy sources as their power resources. However, an alternative transmitter can fulfill power transfer during a separate time period. This may be suitable for mobile loads (Kuffler & Yoshikami, 1975).

6. **Receivers Quantity:** While the typical topology for WPT systems typically focuses on a single receiver, there are alternative configurations that have been developed to accommodate multiple loads. Therefore, it is feasible for several recipients to benefit from the advantages of the energy produced by a single sender (Wu et al., 2014).
7. **Stationary/Mobile Receiver:** In certain scenarios, it is necessary for wireless power transfer to accommodate the receiver being positioned randomly prior to initiating the charging process. This highlights the importance of dynamic EV wireless charging (Dahech et al., 2017).
8. **Medium:** While the majority of current WPT products function by having a space between the receiver and power transmitter, this technology has the potential to be utilized in various other mediums, including water, ground, or biological tissue. The choice of medium significantly affects the efficiency, as it is accountable for the losses in power transmission. As an example, a study conducted explored the difference in efficiency between an underwater WPT system and an air-gap system, revealing a decrease of up to 5% in the former (de Man et al., 1960).

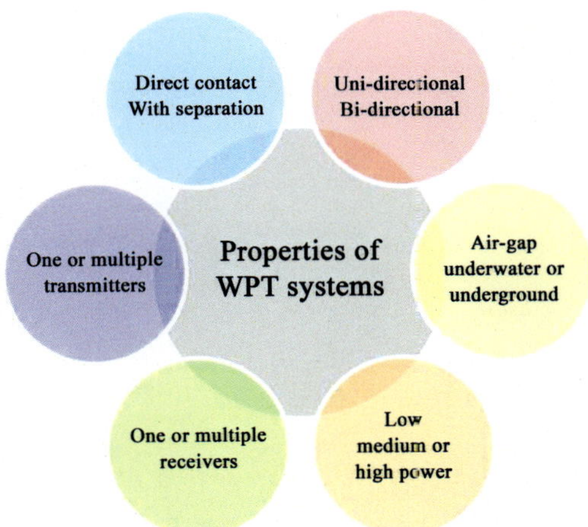

Figure 2.1. *Illustration of WPT systems properties.*

Source: Shubhangi Das, Creative Commons License.

2.2. WIRELESS POWER TECHNOLOGIES

Learning Objectives:

- Understand the classification and operation principles of different wireless power transfer (WPT) technologies.
- Analyze the key components and characteristics of inductive, resonant, capacitive, and magnetic resonance WPT.
- Explore the implementation and advantages, of microwave and optical WPT systems.

Throughout the various experiments and subsequent research, the transmission of power without wires is made possible by the movement of an electromagnetic wave from the power source to the recipient. WPT systems exclusively generate an electromagnetic field for power transfer. On the other hand, energy-harvesting techniques utilize the electromagnetic waves produced to transmit information and obtain energy for device power. Therefore, the limitations of energy-harvesting techniques are only applicable to the demands of information transfer, which do not apply to wireless power transfer technologies (Shoki, 2013).

Figure 2.2 shows a visual representation of a WPT system. The power emitter, or antenna, has a maximum dimension of L_{DEV}. The receiver and transmitter are normally located at a specific distance *d*, which is commonly called the gap (Hui et al., 2013). Electromagnetic waves can be characterized by either their wavelength (λ) or their frequency *f*.

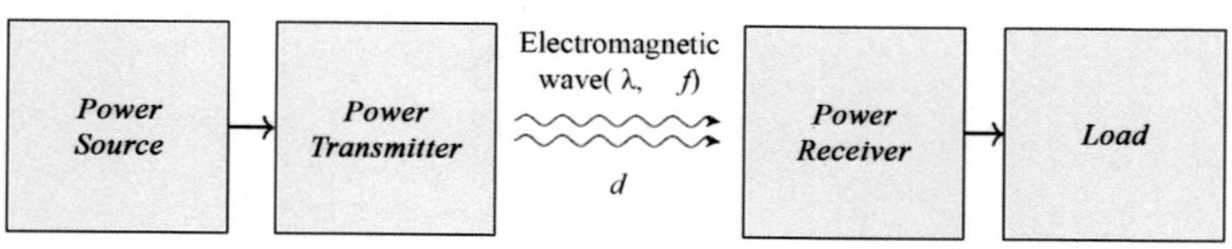

Figure 2.2. *Representation of generic diagram of a WPT system.*

Source: Takehiro Imura, Creative Commons License.

Maxwell's equations define the characteristics of an electromagnetic wave. When certain conditions are met, these complex equations can be made simpler, resulting in the operation of the system in the far-field and near-field. Next, both scenarios will be discussed (Song et al., 2021).

- **Non-Radiative Propagation or Near-Field Operation:** In order for this kind of situation to function, three requirements must be met. They are:

- The wavelength λ is substantially larger than the size of the transmitter element, L_{DEV}.
- The distance between the receiver and source of energy is smaller than the wavelength λ.
- The distance between receiver and transmitter is significantly less than the ratio of the square of the device length to the wavelength $(L^2_{DEV}) / \lambda$ (Courjon et al., 1993).

• **Radiative Propagation or Far-Field Operation:** This is based on the electromagnetic wave's electric field. Here, the circumstances are:

- Ther receiver and energy emitter are farther apart than the wavelength λ.
- The wavelength λ is less than 10 times less than the size of the transmitter element L_{DEV} (Sarkar & Taaghol, 1999).

Within each scenario, a collection of WPT technologies is outlined in the chart provided. There are different methods of wireless power transfer that can be used in near-field operation, including inductive, resonant, and capacitive techniques (Bayliss & Turkel, 1982). Other far-field technologies that can be used for wireless power transfer are those that are based on microwaves or optical technology. Between near-field and far-field technologies, Strongly Coupled Magnetic Resonance Systems represent an intermediate configuration with different modes of operation. Understanding the categorization of WPT technologies is crucial for an accurate analysis of electrical systems. It is important to consider the simplification of Maxwell's equation through the application of Kirchhoff's Law in near-field operation. However, in far-field operation, a more comprehensive approach is required, involving RF analysis and optics-based equations (Shrestha et al., 2011). Next, all the WPT technologies mentioned above will be discussed (Figure 2.3).

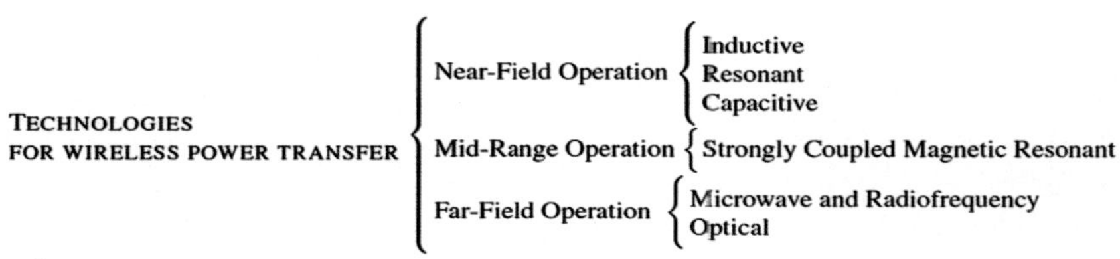

Figure 2.3. *Illustration of classification of WPT technologies.*

Source: Takehiro Imura, Creative Commons License.

2.2.1. Inductive WPT

The electromagnetic wave's magnetic field is used to achieve inductive WPT. Ampère's Law and Faraday's Law, which describe an interaction of magnetic and electrical activity, explain the operation principle (Costanzo et al., 2017).

A current-carrying wire creates a magnetic field surrounding it, according to Ampère's Law. The strength and direction of the magnetic field are determined by the wire's topology. Ampère's Law specifically states that:

$$\oint \overline{H} dl = I \tag{2.1}$$

The magnetic field strength $\overline{H}$ is determined by the electric current (I) and the differential element (dl) along its path. The magnetic field's strength variation aligns with the wire's current frequency due to this physical phenomenon. Figure 2.4 shows the magnetic fields of various structures commonly used in inductive wireless chargers (Mayordomo et al., 2013).

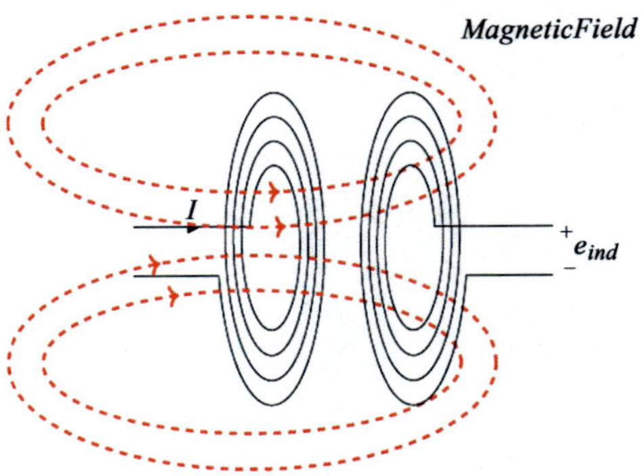

Figure 2.4. *Illustration of induced voltage caused by a changing magnetic field.*

Source: Johnson I. Agbinya, Creative Commons License.

As demonstrated, coils can enhance the concentration of the magnetic field in their defined area more effectively than a basic wire (Le et al., 2008).

According to Ampère's Law, a coil carrying a time-varying current generates a time-varying current magnetic field around it. A voltage (e_{ind}) is induced in the terminals of a separate coil if that time-varying magnetic field passes through it. According to Faraday's Law, this effect is as follows:

$$e_{ind} = -\frac{d\phi}{dt} \tag{2.2}$$

where; φ is the magnetic field flux flowing through the coil-limited region.

Many magnetic-based technologies utilized in wireless power transfer are based on the combination of these two phenomena. The inductive wireless power transfer technology involves the use of two coils, known as the secondary and primary coils. Here is a diagram shown in Figure 2.5. A generator is required to create a time-varying current in the primary coil. The magnetic field generated needs to pass through the secondary coil, where the load (R_L) is connected, in order to power or charge it. In most cases, there are electronic components placed between the primary coil and generator. Furthermore, there are multiple electrical systems that establish a connection between the load and secondary coil. These extra components can improve wireless power transfer efficiency and are discussed in the section that follows (Kuperman, 2020).

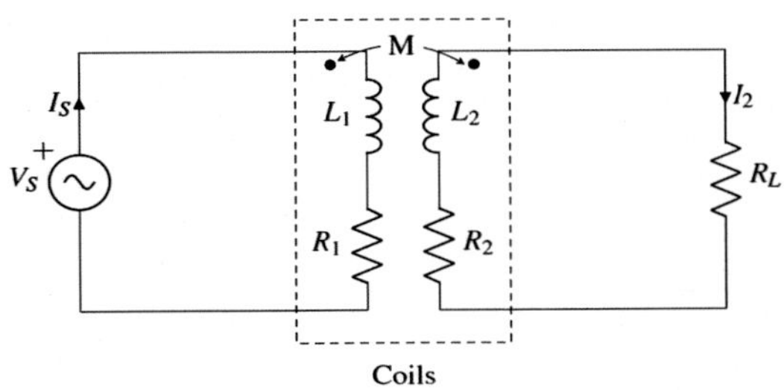

Figure 2.5. *Equivalent circuit of inductive WPT.*

Source: Yiming Zhang, Creative Commons License.

It is advisable to generate an induced voltage that is maximized. According to Faraday's Law, the voltage induced is directly related to how quickly the magnetic field changes in the secondary coil. When a coil is exposed to two magnetic fields with the same strength but varying frequencies at different moments, it will generate two distinct induced voltages. At the maximum frequency of the magnetic field traveling through the coil, there will be a higher induced voltage. Thus, it is preferable to enhance the secondary coil's magnetic flux fluctuation (Monti et al., 2020).

Therefore, it is important for an inductive-based wireless power transfer system to meet these two conditions:

- The majority of the magnetic field produced by the primary coil passes through the secondary coil; and

- The magnetic field frequency used in wireless power transfer is maximized to enable near-field operation (Regensburger et al., 2020).

One important consideration is that larger coils are initially preferred on the secondary side. However, the application places certain restrictions on the size, weight, and cost of this component. This limitation is obvious in biomedical applications. For EV applications, there are certain limitations on the size of the coils due to the structures they need to fit into, and the cost of materials involved. It is important to understand that conventional inductive wireless power transfer (WPT) is not required for the support of wireless power transfer (WPT) in electric vehicles (EVs). Rather, it makes use of cutting-edge technologies built around magnetic WPT (Shinohara, 2020).

Increasing the frequency of the electrical current in the primary coil is the key strategy for enhancing wireless power transfer (WPT) in inductive systems, as it leads to a larger rate of flux change, which is beneficial for inductive WPT. Consequently, the magnetic field's frequency will rise, leading to an amplified rate of flux change (Dai & Ludois, 2015).

In magnetic-based WPT systems, the function of power converters is to raise the operating frequency. There are two commercial technologies that utilize inductive wireless power transfer: Qi and Radio Frequency Identification (RFID).

2.2.2. Magnetic Resonance WPT

Resonant wireless power transfer, also known as magnetic resonance, can be seen as an advancement over inductive wireless power transfer. This method involves operating the electrical system under resonant conditions, which enhances its performance. To meet this requirement, structures made up of reactive components such as capacitors or extra coils are attached to the two coils. The term "compensation networks" is often used to describe these arrangements. Figure 2.6 shows an example of a common resonant wireless power transmission system (Hui, 2016).

One of the fundamental compensation topologies utilizes a single capacitor that can be connected either in series or parallel to both the primary and secondary components. These networks are commonly known as mono-resonant compensation topologies.

Additionally, there are alternative compensation topologies that are more intricate. These topologies are often known as multi-resonant compensation topologies. The technology used for wireless charging in electric vehicles is based on resonance (Mou et al., 2019).

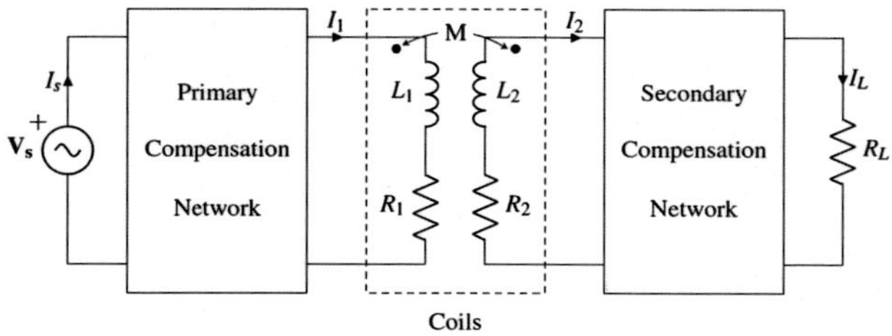

Figure 2.6. *Illustration of magnetic resonance wireless chargers with compensation networks.*

Source: Qiang Ni, Creative Commons License.

2.2.3. Capacitive WPT

Unlike previous near-field WPT techniques, capacitive WPT utilizes the electrical field to achieve its goals. When dealing with capacitive wireless power transfer, it is necessary to incorporate two metallic plates into both the power emitter and the receiver. As seen in Figure 2.7, each component of the setup involves attaching a single plate to the ends of the wires linking the load and the power supply (Lecluyse et al., 2021).

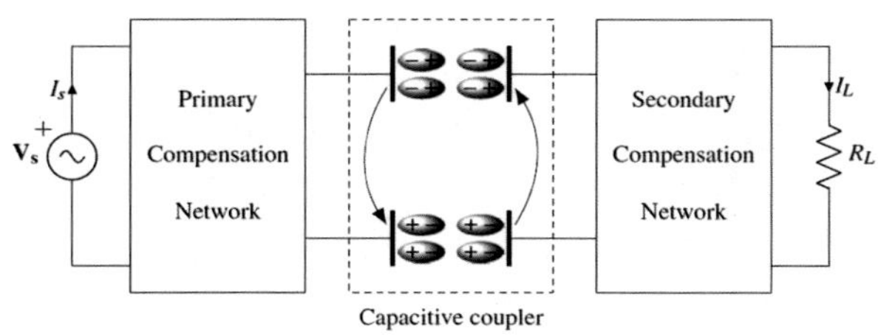

Figure 2.7. *Illustration of capacitive wireless chargers with compensation networks.*

Source: Yiming Zhang, Creative Commons License.

The emitter's two ends have plates that are parallel to the matching plates on the receiver. The electrical circuit closes when the two pairs of plates are near each other, acting as capacitors. The terms "forward and return capacitors" are frequently used to describe these capacitors.

In the given scenario, capacitive coupling occurs. Consequently, the interaction between the plates leads to the generation of an electric field, which induces an electrical current in the power receiver. Just like magnetic-based wireless power transfer, the amount of current induced in the power receiver depends on how quickly the electric field flux changes between the two pairs of plates. Using grid power, the operational frequency is increased to improve this rate. Power converters are used to generate for this purpose (Erel et al., 2021).

Unlike magnetic-based techniques, capacitive wireless power transfer can transfer power even in the presence of metallic objects. In wireless power transfer, the impact of intermediate metallic objects on the process is negligible in various applications. The losses are negligible and the objects themselves do not become excessively hot (Kracek & Svanda, 2018).

One notable benefit is that the electrical field is limited to the area between the two plates. Consequently, the subject remains closely tied to this area, despite previous research focusing on the magnetic field in WPT techniques. Currently, it appears that resonant technology has reached a level of development that can effectively facilitate a greater degree of power transfer.

2.2.4. Strongly Coupled Magnetic Resonance WPT

In the relevant literature, the terms "strongly coupled magnetic resonance" (SCMR) and "WPT based on magnetically coupled resonance" are used interchangeably.

SCMR systems are categorized as mid-range wireless power transfer (WPT) systems due to their unique characteristics that do not fit only into the near-field or far-field functioning categories, distinguishing them from earlier technologies. While the first condition, $L_{DEV} \ll \lambda$, is met, the other two conditions required for near-field operation are not satisfied. Specifically, for mid-range operations, it has also been confirmed that the distance is between 1 and 10 times the length of the device $1 < d < 10*L_{DEV}$. However, the L_{DEV} can be significantly greater than the distance between the load and power source (Wei et al., 2014).

Researchers from the Massachusetts Institute of Technology conducted the first significant experiment with SCMR technology in 2007. They could power a light bulb in this experiment with no cords, and the component was placed two meters distant from the power source. In this case, the power transmission efficiency was close to 40%.

The primary characteristics of SCMR circuits were established by this experiment. They are:

1. **4-Coil Topology:** A frequently used design for SCMR circuits involves four interconnected coils, with two coils in the power

source (the driver and the transmitter) and two coils in the load (the receiver and the load). All the components are magnetically connected, without the use of any conductive materials. The topology is displayed in Figure 2.8. The coupling coefficient between the coils is also displayed in the figure.

2. **The Self-Resonant Nature of the Intermediate Coils:** The free resonant frequency of the transmitter and receiving coils determines the system's operational frequency. The intermediate coils' parasitic capacitance is used to calculate this value. In consideration of the coils' physical characteristics, the free resonant frequency is often adjusted within the range of 100 kHz to 20 MHz.
3. **The Load Loop and Driver Compensation Topologies:** A basic loop is formed by the coils that are connected to the load and the power source. In order to achieve resonance in the system, it is necessary to use lumped reactive components, as their parasitic capacitance is not significant (Jolani et al., 2015).

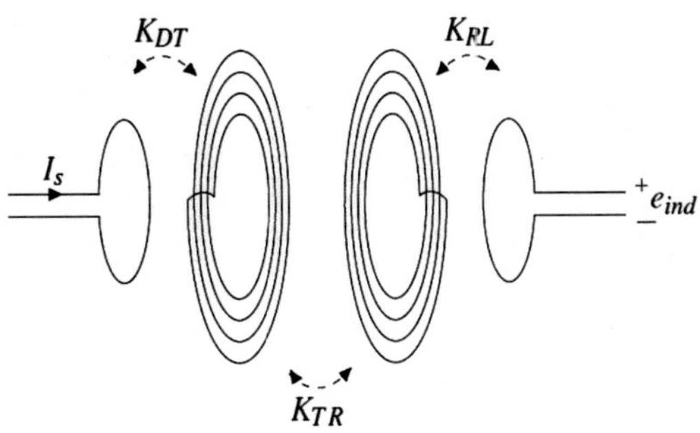

Figure 2.8. *Illustration of four-coil topology in SCMR systems.*

Source: Soares, F. J.; Creative Commons License.

The effectiveness of SCMR systems in the mid-range scenario relies on the utilization of either a 4-coil or indirect-link topology. To gain a better understanding of the properties, one can make use of a two-coupled coil system that is equivalent. In this model, the coupling coefficient is expressed as a simplified equation involving KDT * KRL/KTR. K_{TR} indicates the coupling coefficient between the transmitter and receiver, K_{RL} indicates the coupling coefficient between the load and the receiver, and K_{DT} indicates the coupling coefficient between the drivers and the transmitter (Barreto et al., 2020).

The degree to which one coil's magnetic flux is used by another to produce a voltage is indicated by the coupling coefficient. In order to maximize wireless power transfer, coupling coefficients close to 1 must be targeted. Furthermore, it is crucial to fulfill additional system conditions, such as ensuring the proper separation between the power supply and the load. Typically, when two interconnected coils are positioned at a greater distance from each other, the coupling coefficient tends to diminish.

Upon analyzing the equivalent coupling coefficient, it is evident that this parameter has the potential to reach a value of 1, even when the coupling coefficient between the transmitter and receiver coils is rather low. Proper adjustment of the other two coupling coefficients, K_{DT} and K_{TR}, is necessary to achieve this. It is important to note that some SCMR control algorithms can change the distance between the transmitter coil and the driver loop, which changes the coupling coefficient KDT. Two major goals drive SCMR system control algorithms: achieving the most efficient power transfer to the load and maximizing the process efficiency. In order to accomplish these objectives, the system makes adjustments to certain parameters (Zhou et al., 2018).

The solutions mentioned earlier include modifying the operational frequency, integrating flexible matching networks, and altering the distance between the transmitter coils and drivers. When examining these three approaches, it is crucial to take into account the occurrences of frequency bifurcation and frequency splitting. Resonant structures often experience frequency bifurcation.

2.2.5. Microwave Power Transfer

When discussing WPT based on microwave energy transmission in a far-field setting, the term "microwave power transfer" (MPT) is used. With a few small adjustments, the process can also be used to radiofrequency (RF) signals. Indeed, RF and MPT power transmission are occasionally combined into wave-based WPT (Huang & Zhou, 2015).

Figure 2.9 shows the generic structure of an MPT. A magnetron generates a microwave signal using a high-voltage DC generator. After that, the microwave passes through the antenna. The signal is processed by the receiver using a rectenna, which changes the microwave signal into a DC signal. Ultimately, the DC signal is received by the power electronics equipment.

It is evident that the key components of the MPT are:

1. **Magnetron:** It is an oscillator that is this vacuum tube. Without any amplification, the magnetron produces a microwave signal from a DC excitation. The magnetron's size are correlated with the frequency of the generated signal.

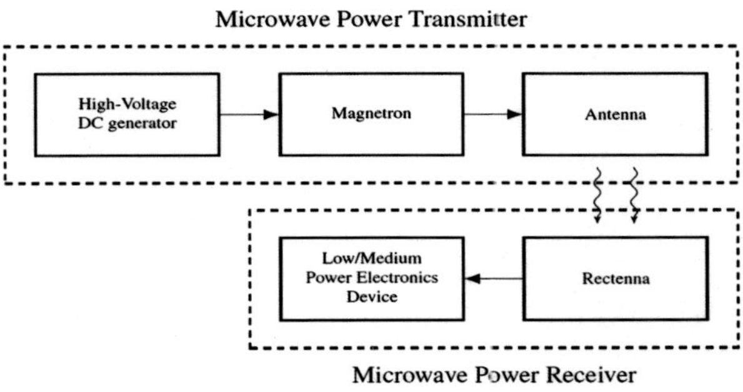

Figure 2.9. *Illustration of generic diagram of a MPT system.*

Source: Francesco Marra, Creative Commons License.

2. **Antenna:** This is equivalent to the electrical part that can produce an electromagnetic wave and send it to the receiver. The signal is produced by the most basic antennas in an isotropic manner, meaning that its power is constant across all spatial directions. As an alternative, beam shaping can be achieved by varying the current phase difference using a range of antennas (Khang et al., 2017).
3. **Rectenna:** This refers to a combination of an antenna and a rectifying circuit integrated into a single component. The design was created by Charles Brown in 1960 to function with a 2.45-GHz signal.

An experiment on MPT was conducted in 1964, which is considered being the first relevant one. William Brown and his team successfully demonstrated the use of microwave power transmission to power a model aircraft. In 1968, Peter Glaser introduced the concept of Space Solar Power Satellite (SSPS), which revolutionized the field of energy transmission. This research guideline focuses on the transfer of energy from a satellite to the Earth by reflecting energy from the Sun. In 1975, they achieved a significant milestone by successfully transferring 450 kW of power to a receiver located 1 mile away from the transmitter, due to MPT (Zhu et al., 2020).

Small aircraft operating at altitudes of up to 21 kilometers were fueled during the 1980s as part of the Stationary High Altitude Relay Platform (SHARP) experiment. The aircraft offered telecommunication services from this height. It was possible to transfer 500 kW of power using a 5.8 GHz transmission. The Kyoto University professor Shinohara's group has been working on developing microwave-based wireless chargers for electric vehicles (EVs), with some of their prototypes reaching up to 84% efficiency. Currently, some commercial solutions use MPT in home environments. Power Cast is one of these. A 915-

MHz signal produced by a power cast source can transmit up to 3 W in a directed beam or up to 1 W in an isotropic one. Cota is an Ossia product designed for WPT that also supports radiofrequency transmission. To identify obstructions or people in the transmission path, it uses advanced algorithms (Strassner & Chang, 2013).

Its functionality also requires a Bluetooth channel. WattUp was created by Energous for WPT between 2.4 and 5 GHz. The main characteristic of WattUp is its utilization of a variety of antennas in the transmitter to regulate beamforming and improve the efficiency of power transfer. Additionally, the receiver features several antennas to increase its energy absorption from various electromagnetic waves. The maximum efficiency of 70% can be achieved with a distance of up to 5 m. Low-power systems are evident in the commercial applications. MPT becomes hazardous at greater power levels.

2.2.6. Optical WPT

Power transfer in optical WPT systems is achieved through the utilization of electromagnetic waves with a frequency in the terahertz (THz) range. For this wave to be efficient, the power transmitter and its receiver must have a clear line of sight. This means that there should be no obstacles in between, as the wave cannot pass through them. If this requirement is met, the distance between the two points can extend to several kilometers. This was the situation with the initial experiments on optical wireless power transfer. An initial experiment was conducted to test the capabilities of this technology, involving the powering of a Mini Rover over a distance of up to 280 m. The laser's output power was 5 W, while the vehicle only required 1 W.

NASA carried out an additional experiment in 2003, three years later, using a 1-kW laser beam to remotely power a 6-W drone. Despite several UAV experiments that have been reported in the literature, the transmission's power and efficiency stayed lower. Furthermore, a regulated optical WPT can have the significant advantage of not interfering with other systems due to its directivity (Nguyen, 2023).

Currently, it is challenging for electronics to easily modify the direction of the beam. However, this task can be accomplished using passive elements such as lenses. One notable drawback is the low efficiency of these systems, which typically hovers around 25%. Advancements in laser technology and photovoltaic cells show promising potential for future improvements in this area.

Figure 2.10 displays a diagram of a typical optical wireless power transfer system. Based on this setup, the transmitter's laser diode is regulated by the generator's current mode. The direction of the power transfer is managed by a beam director. The power converters in the receiver convert the received light into power using a photovoltaic cell. The DC power is subsequently utilized to provide power to a load or charge a battery (Putra et al., 2020).

Optical WPT systems, particularly those operating in the infrared, are safe to use in human surroundings for low-power applications. This has led to the commercialization of home-use optical WPT products. An example of an infrared optical WPT system for low-power devices is Wi-Charge. In its standard setup, the transmitter is often mounted on a room's ceiling. It has advanced electronics that allow it to adjust the beam to match the receiver. The specifications state that they can charge up to 3 W in a kilometer (Wong et al., 2022).

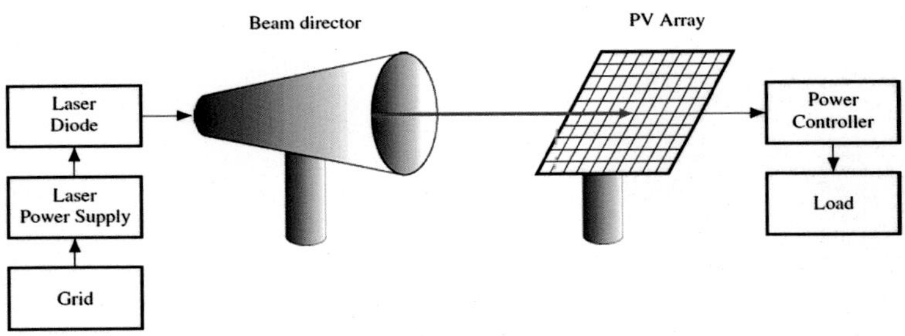

Figure 2.10. *Illustration of generic diagram of an optical WPT system.*

Source: Fabian Wirth, Creative Commons License.

WPT uses a variety of technologies. The principles of near-field, mid-range, and far-field technologies have been covered in the earlier sections. Table 2.1 lists these technologies' principal benefits and drawbacks (Tang et al., 2020).

2.3. ASSOCIATED TECHNOLOGIES

Learning Objectives:

- Understand the receiver localization techniques for efficient WPT performance.
- Understand the concept of foreign object detection and data communication.

Real WPT products are complex systems that include extra electronics and processes to guarantee that the power transfer process is carried out in a controlled manner, in addition to the technology for transferring power via the medium.

The implemented controls aim for an electrical and magnetic performance that is both practicable and safe. The system must also be interconnected with other infrastructures or systems. The most important technologies found in actual, commercial WPT products are described in subsections (Tomkiewicz et al., 2010).

2.3.1. Data Communication

It is necessary to charge devices and their batteries in a certain way to maintain certain electrical magnitudes within a specified time frame. For example, certain EV batteries need a constant current throughout a certain portion of the charging process, and the voltage is then maintained at that level. The components of the wireless transmitter must modify how the power signal is generated based on the battery's condition in order to adhere to these limitations (Cunha et al., 2016).

Table 2.1. *Properties of WPT Technologies*

WPT Technologies		**Key Advantages**	**Potential Disadvantages**
Near field	Capacitive	• Transfer of medium power up to several kilowatts. • Transfer between the receiver and transmitter, even when metal objects are present. • It uses inexpensive metal plates, which are used to transport power. • Appropriate for tiny applications with a 10 cm or less gap. • Limitation of the electric field to the space between the plates eliminates the requirement for EMI management.	• Efficiency in the 70–80% range. • Power transfer is strongly influenced by the distance. • Smaller distance (up to 10 cm).

	Inductive	• Basic implementation suitable for low power applications. • Galvanic isolation is provided. • Basic control.	• Efficiency is significantly limited at a range of 20%. • Minimal distance for transmission (up to a few centimeters). • Importance of regulating electromagnetic emissions. • The effectiveness of wireless power transfer is greatly influenced by the presence of obstacles, particularly metallic ones, near the transmitter and receiver.
	Resonant	• Efficient power transfer of several kilowatts. • Established technology with practical uses in the market. • Power transfer remains possible despite slight changes in the receiver's position, yet with a decrease in efficiency. • Galvanic isolation provided.	• Costly components used in high-power applications. • The effectiveness of wireless power transfer is heavily influenced by the presence of obstacles, particularly metallic ones, near the transmitter and receiver. • Importance of regulating electromagnetic emissions.

Mid-range	SCMR	• Effective transmission range of 2–3 meters. • Suitable for mobile devices. • Appropriate for low and medium power applications.	• The efficiency is less than 50%. • The implementation of the coils and control algorithms can be quite complex. • The receiver and transmitter both have large and bulky components.
Far field	Microwave	• Effective range of several kilometers. • The microwave beam can be dynamically configured to adapt the power transfer to mobile loads. • There is a significant potential to transfer several kilowatts of power.	• Inefficiency is a concern. Only a small fraction, specifically less than 10%, is achieved for high power applications with long gaps. • Challenging execution. • Potential risks associated with the operation involving living beings between the power transmitter and the receiver. • Significant dimensions of the receiver.

	Optical	• Extended effective distance of several km. • It enables the light beam to be dynamically configured to adjust power transfer for moving loads. • Possibility of moving several kilowatts of electricity smaller transmitter.	• Low efficiency (around 20%). • There must be a clear path of sight between the transmitter and the receiver; obstructions are not permitted. • If the infrared region is not used, there is an unsafe operating environment for living things between the power transmitter and the receiver.

Source: Robert Shorten, Creative Commons License.

Although some studies have suggested techniques to determine the battery's condition from measurements made in the transmitter, these techniques are highly dependent on component tolerance. As a result, the power transmitter needs to receive information about the battery. Furthermore, in addition to being electrically controlled, wireless power transfer can also be restricted in its operation. In this way, WPT cannot be started until the user or device has successfully authenticated, according to certain public infrastructures (Papadimitratos & Haas, 2006).

On the other hand, private chargers require users to provide their information for billing purposes. While charging, it may be important to keep the power receiver updated on any changes in energy prices to ensure they are informed about the economic factors involved. Data communication between the power transmitter and receiver supports these procedures. Wireless chargers communicate over a wireless channel to deliver power without a cable (Pahlavan & Levesque, 1994).

Modern wireless chargers use proven cellular or ad hoc wireless technologies (IEEE 802.11 or IEEE 802.16). Certain experimental prototypes of magnetic-resonance chargers suppress the need for a separate communication system by transmitting data through the power coils as well (Godard, 1980). There have been recent studies that have also focused on the security aspects of wireless data transfer for wireless chargers. These have specifically concentrated on how to prevent unauthorized users from accessing WPT. This is accomplished in their proposed magnetic-resonance charger by altering the pulses' frequency and length. Subsequently, the receiver's compensating network capacitor is modified (Kretzmer, 1966).

2.3.2. Receiver Localization

Certain WPT systems enable mobile receivers. The transmitter must produce the electromagnetic wave in accordance with the receiver's current position for these chargers to function effectively. The method used to accomplish this varies based on the WPT technique. Beam forming is the method used by far-field WPT systems to deliver power to the intended location (Noroozi & Sebt, 2015). Certain transmitters are made to function as an array of sources in order to achieve this functionality. The difference in phase of the currents produced by the sources is then used to regulate the beam (Rui & Ho, 2014).

Practice Problem

Tesla, the renowned electric vehicle manufacturer, is seeking to implement wireless power transfer technology for its fleet of electric vehicles. As an engineer at Tesla, you are tasked with selecting the most suitable WPT technology for charging Tesla vehicles. Consider factors such as power transfer efficiency, distance, and safety to justify your choice of WPT technology.

Solutions to Practice Problem

After careful evaluation, Tesla opts for magnetic resonant coupling as the preferred WPT technology for charging its electric vehicles. This choice is driven by several factors. Firstly, magnetic resonant coupling offers high power transfer efficiency, ensuring minimal energy loss during charging, which aligns with Tesla's commitment to energy efficiency. Additionally, its ability to transfer power over longer distances makes it suitable for dynamic EV wireless charging scenarios, where vehicles may not always be parked in fixed positions. Safety is paramount for Tesla, and magnetic resonant coupling provides galvanic isolation, enhancing safety by preventing direct electrical contact between the vehicle and the charging infrastructure. Given these considerations, magnetic resonant coupling emerges as the ideal WPT technology to meet Tesla's charging needs effectively and safely.

As an alternative, near-field WPT methods modify the generated wave levels to provide only the required amount of power to the receiver. Furthermore, a number of transmitters might be ready to go when needed (Ho et al., 2007). This is frequently used in dynamic EV wireless chargers and power mats. To turn on only those transmitters that could create an effective power transfer in these structures, one has to find the receiver. As mentioned in, for magnetic power transfer, the mutual inductance can be a reliable indicator of how close a transmitter coil is to the receiver (Li et al., 2020).

2.3.3. Foreign Object Detection

Certain wireless power transfer technologies performance can be influenced by the presence of objects obstructing the path between the receiver and power transmitter. In RF-based wireless power transfer, these objects become largely insignificant, particularly when considering omnidirectional transmission. With optic wireless power transfer, the power transmission is disrupted by these items, showing a contrasting behavior. The efficiency of magnetic-based power transfer in wireless systems can be significantly impacted when the receiver coil does not receive sufficient magnetic flux (Dai et al., 2020).

Intermediate objects can also lead to unforeseen mishaps, in addition to affecting efficiency. The power transfer in EV wireless chargers involves the induction of eddy currents in objects through the magnetic field. Consequently, the object's temperature rises, potentially resulting in combustion.

It is essential to treat these impacts with utmost seriousness when living organisms are involved. Wi-Charge halts the power transfer when it detects the presence of an intervening object or entity (Lu et al., 2021).

The Foreign Object Detection (FOD) algorithms are designed to detect instances where objects or living persons are present in the area between the receiver and transmitter, causing a disruption in power transfer. In magnetic-based wireless power transfer (WPT), the process involves determining the mutual inductance of the coils. If this is modified, it is presumed that there is an intermediate object present in the conflict zone (Tang et al., 2023). As previously explained, it is important to utilize this reasoning carefully and to determine if changes in this electrical characteristic are caused by external objects or misalignment of the coil can be difficult. The Living Object Protection (LOP) algorithms utilize Foreign Object Detection (FOD) to ensure the safety of Wireless Power Transfer (WPT) systems (Hogeweg et al., 2013).

2.4. ADVANCES IN WPT TECHNOLOGY

Learning Objectives:

- Understand the significance of WPT in electric vehicle charging and SPS technology.
- Recognize the challenges and future prospects of WPT.

Kurs delivered the inaugural demonstration of WPT in 2007. Subsequently, significant advancements have been made in magnetic resonant coupling for its utilization in commercial contexts. In 2008, Intel made alterations to the magnetic resonant coupling by replacing the helix coils with flat coils, which were more convenient to accommodate in a mobile device (Shinohara, 2020).

Kurs founded Witricity Corp. and, in collaboration with the TED Global 2009 conference, showcased Wireless Power Transfer (WPT) technology for portable devices like cell phones. Subsequently, he refined the technology by precisely adjusting interconnected resonators, enabling the simultaneous passage of energy to many coils. This advancement found practical usage in various home and office devices, such as laptops, tablets, and cell phones (Mayordomo et al., 2013).

In 2010, Haier showcased an innovative wireless HDTV that eliminated the need for power cords and signal cables. In recent times, a number of automobile companies such as Audi, Nissan, Rolls-Royce, Toyota, and Mitsubishi have been engaged in the development of wireless power transfer technology to fuel electric or plug-in hybrid vehicles (Adepoju et al., 2022).

In 2011, Rolls-Royce successfully produced an electric version of its Phantom car. With the advancement of WPT technology, electric vehicles can now conveniently charge while parked on the street or in a garage, eliminating the need for complicated power cords. Once fully developed, this technology has the potential to greatly benefit the electric car industry (Sun et al., 2018).

Wireless electricity is a technology that relies on the principles of strongly coupled magnetic resonance. Using two identical resonating coils, a 60-Watt bulb was successfully illuminated at a distance of seven feet from the power source, achieving a power transfer efficiency of 40% (Perera et al., 2017).

The Solar Power Satellite (SPS) is widely regarded as an example of technological advancement. Therefore, the transmission of power wirelessly plays a crucial role in transferring energy from the geo stationary orbit to the ground. Research is still needed for the advancement of power transmission using microwaves, despite its existence for more than four decades. There has been a change in preference for choosing a frequency in the medical and scientific band (ISM). While 2.45 GHz was previously chosen, the recent advancements in C-band RF technologies have made 5.8 GHz a more desirable option (Feng et al., 2020).

Sasaki has determined the most suitable medium, either microwaves or lasers, for their experiment. They plan to conduct a 100-kW class experiment in space before 2020. Testing of power capacities ranging from 2 MW to 200 MW is planned to be conducted by 2030, taking into consideration the anticipated power costs. This scenario would ensure the development of a 1 GW class commercial SPS by the 2030s (Dautov et al., 2020).

SUMMARY

- The chapter titled "Fundamentals of Wireless Power Transfer" offers a thorough examination of the principles, classifications, and technologies related to WPT. The different types of wireless power transfer are classified based on the method of power transfer. These include inductive, magnetic resonance, capacitive, strongly coupled magnetic resonance, microwave, and optical methods.
- Power can be transferred through magnetic fields generated by coils in the process of inductive wireless power transfer. Utilizing resonant inductive coupling, Magnetic Resonance Wireless Power Transfer (WPT) significantly improves both efficiency and range. Capacitive wireless power transfer utilizes electric fields generated between conductive plates to transmit power.
- Exploring the concept of Strongly Coupled Magnetic Resonance Wireless power transfer (WPT) maximizes efficiency and extends the range by optimizing the resonance between the transmitter and receiver. There is a method called Microwave Power Transfer that utilizes microwave radiation to transmit power over long distances. Optical wireless power transfer utilizes laser beams or other light sources to transmit power.
- Other technologies that are related to this field involve data communication to manage power transfer and ensure optimal performance, receiver localization to accurately determine the position of the power receiver, and foreign object detection to prioritize safety and prevent any interference. Recent developments in WPT technology focus on enhancing efficiency, expanding range, ensuring safety, and seamlessly integrating with a wide range of applications and devices.

REVIEW QUESTIONS

1. What are the main classifications of Wireless Power Transfer (WPT) technologies? How do they differ?
2. Explain the working principles of inductive WPT and magnetic resonance WPT. What are their key differences?
3. Discuss the advantages and limitations of inductive WPT, magnetic resonance WPT, capacitive WPT, microwave power transfer, and optical WPT.

4. How does strongly coupled magnetic resonance WPT improve efficiency compared to traditional magnetic resonance WPT? What factors influence its effectiveness?
5. What are the key application areas of WPT beyond power transmission? Provide examples and explain their usage.
6. How does receiver localization advance WPT technology? What are the methods and challenges in localizing receivers accurately?
7. What is foreign object detection in WPT, and why is it important? Describe techniques for detecting foreign objects and preventing hazards.
8. What are potential future developments and research directions in WPT? Discuss emerging trends and challenges.

MULTIPLE CHOICE QUESTIONS

1. What is one of the primary classifications of Wireless Power Transfer (WPT)?

a. Thermal WPT

b. Radioactive WPT

c. Magnetic WPT

d. Electrical WPT

2. Which WPT technology utilizes tightly coupled resonators to enhance efficiency.

a. Inductive WPT

b. Magnetic Resonance WPT

c. Capacitive WPT

d. Microwave Power Transfer

3. In WPT, which technology involves the transfer of power through electric fields?

a. Inductive WPT

b. Capacitive WPT

c. Optical WPT

d. Microwave Power Transfer

4. What is the primary medium used in Microwave Power Transfer?

a. Light

b. Sound

c. Electromagnetic waves

d. Mechanical waves

5. Which application area is NOT commonly associated with WPT?

a. Healthcare

b. Receiver Localization

c. Foreign Object Detection

d. Deep-Sea Mining

6. What is one of the key advantages of Magnetic Resonance WPT over Inductive WPT?

a. Higher efficiency

b. Lower cost

c. Greater range

d. Simplicity of design

Answers to Multiple Questions

1. (c); 2. (b); 3. (b); 4. (c); 5. (d) 6. (a)

REFERENCES

1. Adepoju, W., Bhattacharya, I., Sanyaolu, M., Bima, M. E., Banik, T., Esfahani, E. N., & Abiodun, O. (2022). Critical review of recent advancement in metamaterial design for wireless power transfer. *IEEE Access, 10*, 42699–42726.
2. Barreto, J., Kaddour, A. S., & Georgakopoulos, S. V. (2020). Conformal strongly coupled magnetic resonance systems with extended range. *IEEE Open Journal of Antennas and Propagation, 1*, 264–271.
3. Bayliss, A., & Turkel, E. (1982). Far field boundary conditions for compressible flows. *Journal of Computational Physics, 48*(2), 182–199.
4. Clerckx, B., Zhang, R., Schober, R., Ng, D. W. K., Kim, D. I., & Poor, H. V. (2018). Fundamentals of wireless information and power transfer: From RF energy harvester models to signal and system designs. *IEEE Journal on Selected Areas in Communications, 37*(1), 4–33.
5. Costanzo, A., Dionigi, M., Mastri, F., Mongiardo, M., Monti, G., Russer, J. A., & Tarricone, L. (2017). Conditions for a load-independent operating regime in resonant inductive WPT. *IEEE Transactions on Microwave Theory and Techniques, 65*(4), 1066–1076.
6. Courjon, D., Bainier, C., Girard, C., & Vigoureux, J. M. (1993). Near field

optics and light confinement. *Annalen der Physik, 505*(2), 149–158.

7. Cunha, F., Villas, L., Boukerche, A., Maia, G., Viana, A., Mini, R. A., & Loureiro, A. A. (2016). Data communication in VANETs: Protocols, applications and challenges. *Ad Hoc Networks, 44*, 90–103.
8. Dahech, W., Pätzold, M., Gutiérrez, C. A., & Youssef, N. (2017). A non-stationary mobile-to-mobile channel model allowing for velocity and trajectory variations of the mobile stations. *IEEE Transactions on Wireless Communications, 16*(3), 1987–2000.
9. Dai, J., & Ludois, D. C. (2015). A survey of wireless power transfer and a critical comparison of inductive and capacitive coupling for small gap applications. *IEEE Transactions on Power Electronics, 30*(11), 6017–6029.
10. Dai, Y., Liu, W., Li, H., & Liu, L. (2020). Efficient foreign object detection between PSDs and metro doors via deep neural networks. *IEEE Access, 8*, 46723–46734.
11. Dautov, K., Hashmi, M., Nauryzbayev, G., & Nasimuddin, N. (2020). Recent advancements in defected ground structure-based near-field wireless power transfer systems. *IEEE Access, 8*, 81298–81309.
12. de Man, J. D., Rogosa, D., & Sharpe, M. E. (1960). A medium for the cultivation of lactobacilli. *Journal of Applied Microbiology, 23*(1), 130–135.
13. Detka, K., & Górecki, K. (2022). Wireless power transfer—A review. *Energies, 15*(19), 7236.
14. Feng, H., Tavakoli, R., Onar, O. C., & Pantic, Z. (2020). Advances in high-power wireless charging systems: Overview and design considerations. *IEEE Transactions on Transportation Electrification, 6*(3), 886–919.
15. Godard, D. (1980). Self-recovering equalization and carrier tracking in two-dimensional data communication systems. *IEEE Transactions on Communications, 28*(11), 1867–1875.
16. Ho, K. C., Lu, X., & Kovavisaruch, L. O. (2007). Source localization using TDOA and FDOA measurements in the presence of receiver location errors: Analysis and solution. *IEEE Transactions on Signal Processing, 55*(2), 684–696.
17. Hogeweg, L., Sánchez, C. I., Melendez, J., Maduskar, P., Story, A., Hayward, A., & van Ginneken, B. (2013). Foreign object detection and removal to improve automated analysis of chest radiographs. *Medical Physics, 40*(7), 071–901.
18. Howell, S. B., Nelson, L. A., & Rappaport, S. (2001). An exploration of the paradigm for the 2–3-hour period gap in cataclysmic variables. *The Astrophysical Journal, 550*(2), 897.

19. Huang, K., & Zhou, X. (2015). Cutting the last wires for mobile communications by microwave power transfer. *IEEE Communications Magazine, 53*(6), 86–93.

20. Hui, S. R. (2016). Magnetic resonance for wireless power transfer. *IEEE Power Electronics Magazine, 3*(1), 14–31.

21. Hui, S. Y. R., Zhong, W., & Lee, C. K. (2013). A critical review of recent progress in mid-range wireless power transfer. *IEEE Transactions on Power Electronics, 29*(9), 4500–4511.

22. Jiang, C., Chau, K. T., Liu, C., & Lee, C. H. (2017). An overview of resonant circuits for wireless power transfer. *Energies, 10*(7), 894.

23. Jolani, F., Yu, Y., & Chen, Z. (2015). Enhanced planar wireless power transfer using strongly coupled magnetic resonance. *Electronics Letters, 51*(2), 173–175.

24. Khang, S. T., Lee, D. J., Hwang, I. J., Yeo, T. D., & Yu, J. W. (2017). Microwave power transfer with optimal number of rectenna arrays for midrange applications. *IEEE Antennas and Wireless Propagation Letters, 17*(1), 155–159.

25. Kracek, J., & Svanda, M. (2018). Analysis of capacitive wireless power transfer. *IEEE Access, 7*, 26678–26683.

26. Kretzmer, E. (1966). Generalization of a technique for binary data communication. *IEEE Transactions on Communication Technology, 14*(1), 67–68.

27. Kuffler, S. W., & Yoshikami, D. O. J. U. (1975). The number of transmitter molecules in a quantum: An estimate from iontophoretic application of acetylcholine at the neuromuscular synapse. *The Journal of Physiology, 251*(2), 465–482.

28. Kuperman, A. (2020). Additional two-capacitor basic compensation topologies for resonant inductive WPT links. *IEEE Transactions on Power Delivery, 35*(5), 2568–2570.

29. Le, T., Mayaram, K., & Fiez, T. (2008). Efficient far-field radio frequency energy harvesting for passively powered sensor networks. *IEEE Journal of Solid-State Circuits, 43*(5), 1287–1302.

30. Lecluyse, C., Minnaert, B., & Kleemann, M. (2021). A review of the current state of technology of capacitive wireless power transfer. *Energies, 14*(18), 5862.

31. Li, Z., Tian, Z., Wang, Z., & Zhang, Z. (2020). Multipath-assisted indoor localization using a single receiver. *IEEE Sensors Journal, 21*(1), 692–705.

32. Lu, J., Zhu, G., & Mi, C. C. (2021). Foreign object detection in wireless power transfer systems. *IEEE Transactions on Industry Applications, 58*(1), 1340–1354.
33. Lu, X., Niyato, D., Wang, P., Kim, D. I., & Han, Z. (2015). Wireless charger networking for mobile devices: Fundamentals, standards, and applications. *IEEE Wireless Communications, 22*(2), 126–135.
34. Mahesh, A., Chokkalingam, B., & Mihet-Popa, L. (2021). Inductive wireless power transfer charging for electric vehicles–a review. *IEEE Access, 9*, 137667–137713.
35. Mayordomo, I., Dräger, T., Spies, P., Bernhard, J., & Pflaum, A. (2013). An overview of technical challenges and advances of inductive wireless power transmission. *Proceedings of the IEEE, 101*(6), 1302–1311.
36. Monti, G., Mongiardo, M., Minnaert, B., Costanzo, A., & Tarricone, L. (2020). Optimal terminations for a single-input multiple-output resonant inductive WPT link. *Energies, 13*(19), 5157.
37. Mou, X., Gladwin, D. T., Zhao, R., & Sun, H. (2019). Survey on magnetic resonant coupling wireless power transfer technology for electric vehicle charging. *IET Power Electronics, 12*(12), 3005–3020.
38. Nguyen, D. H. (2023). Dynamic optical wireless power transfer for electric vehicles. *IEEE Access, 11*, 2787–2795.
39. Noroozi, A., & Sebt, M. A. (2015). Target localization from bistatic range measurements in multi-transmitter multi-receiver passive radar. *IEEE Signal Processing Letters, 22*(12), 2445–2449.
40. Pahlavan, K., & Levesque, A. H. (1994). Wireless data communications. *Proceedings of the IEEE, 82*(9), 1398–1430.
41. Papadimitratos, P., & Haas, Z. J. (2006). Secure data communication in mobile ad hoc networks. *IEEE Journal on Selected Areas in Communications, 24*(2), 343–356.
42. Patil, D., McDonough, M. K., Miller, J. M., Fahimi, B., & Balsara, P. T. (2017). Wireless power transfer for vehicular applications: Overview and challenges. *IEEE Transactions on Transportation Electrification, 4*(1), 3–37.
43. Perera, T. D. P., Jayakody, D. N. K., Sharma, S. K., Chatzinotas, S., & Li, J. (2017). Simultaneous wireless information and power transfer (SWIPT): Recent advances and future challenges. *IEEE Communications Surveys & Tutorials, 20*(1), 264–302.
44. Putra, A. W. S., Kato, H., Adinanta, H., & Maruyama, T. (2020). Optical wireless power transmission to moving objects using Galvano mirror. In

Free-Space Laser Communications XXXII (Vol. 11272, pp. 314–322).

45. Raso, J., Manas, P., Pagan, R., & Sala, F. J. (1999). Influence of different factors on the output power transferred into medium by ultrasound. *Ultrasonics Sonochemistry, 5*(4), 157–162.
46. Regensburger, B., Sinha, S., Kumar, A., Maji, S., & Afridi, K. K. (2020). High-performance multi-MHz capacitive wireless power transfer system for EV charging utilizing interleaved-foil coupled inductors. *IEEE Journal of Emerging and Selected Topics in Power Electronics, 10*(1), 35–51.
47. Rui, L., & Ho, K. C. (2014). Elliptic localization: Performance study and optimum receiver placement. *IEEE Transactions on Signal Processing, 62*(18), 4673–4688.
48. Sarkar, T. K., & Taaghol, A. (1999). Near-field to near/far-field transformation for arbitrary near-field geometry utilizing an equivalent electric current and MoM. *IEEE Transactions on Antennas and Propagation, 47*(3), 566–573.
49. Shevchenko, V., Husev, O., Strzelecki, R., Pakhaliuk, B., Poliakov, N., & Strzelecka, N. (2019). Compensation topologies in IPT systems: Standards, requirements, classification, analysis, comparison and application. *IEEE Access, 7*, 120559–120580.
50. Shinohara, N. (2020). Trends in wireless power transfer: WPT technology for energy harvesting, millimeter-wave/THz rectennas, MIMO-WPT, and advances in near-field WPT applications. *IEEE Microwave Magazine, 22*(1), 46–59.
51. Shoki, H. (2013). Issues and initiatives for practical deployment of wireless power transfer technologies in Japan. *Proceedings of the IEEE, 101*(6), 1312–1320.
52. Shrestha, B., Elsherbeni, A., & Ukkonen, L. (2011). UHF RFID reader antenna for near-field and far-field operations. *IEEE Antennas and Wireless Propagation Letters, 10*, 1274–1277.
53. Song, M., Jayathurathnage, P., Zanganeh, E., Krasikova, M., Smirnov, P., Belov, P., & Krasnok, A. (2021). Wireless power transfer based on novel physical concepts. *Nature Electronics, 4*(10), 707–716.
54. Strassner, B., & Chang, K. (2013). Microwave power transmission: Historical milestones and system components. *Proceedings of the IEEE, 101*(6), 1379–1396.
55. Sun, L., Ma, D., & Tang, H. (2018). A review of recent trends in wireless power transfer technology and its applications in electric vehicle wireless charging. *Renewable and Sustainable Energy Reviews, 91*, 490–503.

56. Tang, J., Matsunaga, K., & Miyamoto, T. (2020). Numerical analysis of power generation characteristics in beam irradiation control of indoor OWPT system. *Optical Review, 27*, 170–176.

57. Tang, J., Zhou, H., Wang, T., Jin, Z., Wang, Y., & Wang, X. (2023). Cascaded foreign object detection in manufacturing processes using convolutional neural networks and synthetic data generation methodology. *Journal of Intelligent Manufacturing, 34*(7), 2925–2941.

58. Tomkiewicz, S. M., Fuller, M. R., Kie, J. G., & Bates, K. K. (2010). Global positioning system and associated technologies in animal behavior and ecological research. *Philosophical Transactions of the Royal Society B: Biological Sciences, 365*(1550), 2163–2176.

59. Wei, X., Wang, Z., & Dai, H. (2014). A critical review of wireless power transfer via strongly coupled magnetic resonances. *Energies, 7*(7), 4316–4341.

60. Wong, Y. L., Shibui, S., Koga, M., Hayashi, S., & Uchida, S. (2022). Optical wireless power transmission using a GaInP power converter cell under high-power 635 nm laser irradiation of 53.5 W/cm². *Energies, 15*(10), 3690.

61. Wu, Z., Li, S., Yuan, G., Lei, D., & Wang, Z. (2014). Three-dimensional numerical study of heat transfer characteristics of parabolic trough receiver. *Applied Energy, 113*, 902–911.

62. Zeng, Y., Clerckx, B., & Zhang, R. (2017). Communications and signals design for wireless power transmission. *IEEE Transactions on Communications, 65*(5), 2264–2290.

63. Zhang, Z., Pang, H., Georgiadis, A., & Cecati, C. (2018). Wireless power transfer—An overview. *IEEE Transactions on Industrial Electronics, 66*(2), 1044–1058.

64. Zhou, W., Sandeep, S., Wu, P., Yang, P., Yu, W., & Huang, S. Y. (2018). A wideband strongly coupled magnetic resonance wireless power transfer system and its circuit analysis. *IEEE Microwave and Wireless Components Letters, 28*(12), 1152–1154.

65. Zhu, X., Jin, K., Hui, Q., Gong, W., & Mao, D. (2020). Long-range wireless microwave power transmission: A review of recent progress. *IEEE Journal of Emerging and Selected Topics in Power Electronics, 9*(4), 4932–4946.

CHAPTER 3

WIRELESS CHARGERS FOR ELECTRIC VEHICLES

LEARNING OBJECTIVES

At the end of this chapter, readers will be able to:

1. Understand the classification of electric vehicles based on their energy sources.
2. Compare and contrast the battery characteristics of different types of electric vehicles.
3. Explore the evolution of battery technology in electric vehicles, focusing on Lithium-ion batteries.
4. Explore the role and functions of a Battery Management System (BMS) in electric vehicles.
5. Understand the different charging modes and operation modes in electric vehicle charging infrastructure, including wireless charging.
6. Identify key players and technologies in the wireless charging market.
7. Describe the generic structure and operation of EV wireless charger.

INTRODUCTORY EXAMPLE

Smart Electric Vehicles Inc. is a leading manufacturer of electric vehicles (EVs) striving to enhance the charging experience for its customers. Recognizing the growing demand for convenient and efficient charging solutions, Smart Electric Vehicles Inc. has collaborated with WiTricity, a pioneer in wireless power transfer (WPT) technology.

As part of their collaboration, Smart Electric Vehicles Inc. integrates WiTricity's wireless charging systems into their latest models, offering customers the flexibility to charge their EVs without the hassle of cords and plugs. By leveraging WiTricity's expertise in WPT, Smart Electric Vehicles Inc. aims to revolutionize the EV charging landscape, providing users with seamless charging experiences at home, work, and public charging stations.

With WiTricity's innovative technology, drivers can simply park their vehicles over a wireless charging pad, eliminating the need for manual connection and ensuring efficient energy transfer. This partnership between Smart Electric Vehicles Inc. and WiTricity exemplifies the industry's commitment to advancing EV charging infrastructure and fostering a sustainable transportation ecosystem.

UNIT INTRODUCTION

Over the past decade, there has been a notable surge in the advancements of wireless charging systems for electric vehicles. Due to the global phenomenon of rising urbanization, cities are seeking to transition away from gasoline and diesel-fueled vehicles in order to promote the creation of cleaner urban environments. Another factor in this is the increasing efficiency and affordability of electric vehicles. There should not be a need for owners to plug in their cars at all if wireless charging devices are correctly installed in them and placed around cities and their residences (Ahmad et al., 2017).

Drivers can conveniently park their vehicles over a coil that is either placed on the ground or buried beneath it. However, accepting this technology also holds the potential to address various practical and perceived issues. One aspect to consider is that the younger generation of today has a strong desire for freedom and independence. There is a strong inclination to make the world easily accessible and to maintain connectivity while on the move. Not too long ago, both urban and rural areas were filled with telephone boxes, and it was quite common to wait in line to make a phone call while being connected to a wired receiver (Mahesh et al., 2021). Currently, relying solely on tethering for various communication requirements is impractical, especially for basic internet or communication purposes. The inconvenience of having to wait in line just to make a phone call would likely drive most customers to switch to a different network provider. The comparison for electric vehicles is clear. In the future, plug-in charging will likely be replaced by high-power fast charging, as it becomes more necessary and prevalent. In future cities, the need to wait in line for power would ideally be eliminated with the implementation of efficient wireless charging systems (Zhang et al., 2016).

Currently, vehicles are typically charged either at home or at the owner's place of business, especially in the case of fleets. In future, there will be an increased availability of charging options in various locations, such as workplaces, stores, streets, and popular destinations. By installing wireless electric vehicle charging points at these places, it is possible to improve staff and customer loyalty, attract new customers, and encourage the adoption of wirelessly charged automobiles in densely populated areas, resulting in a decrease in air pollution. In an ideal scenario, vehicles could charge effortlessly at any parking spot, without the need for EV owners to worry about connecting to the grid. This seamless process would occur automatically (Amjad et al., 2022).

More practical and effective charging options are becoming more and more necessary as the demand for EVs rises. With wireless chargers for EVs, consumers can expect a hassle-free and seamless charging experience, doing away with the annoyance that comes with regular plug-in chargers. This chapter examines a number of topics related to EV wireless charging technology, such as market views, standards, battery technology, charging modes, varieties of EVs, and the general design of EV wireless chargers. Comprehending these foundational concepts is crucial in order to appreciate the dynamic terrain of electric vehicle charging infrastructure and its consequences for the automotive sector and other sectors (Mohamed et al., 2022).

3.1. ELECTRIC VEHICLE TYPES

Learning Objectives:

- Understand the classification and battery characteristics of electric vehicles.

Given the pressing environmental concerns, there is a significant focus on exploring innovative developments in the sustainable transportation sector. According to this theory, electric cars should significantly lower greenhouse gas emissions, which will lead to a more sustainable and wholesome living environment. Several pieces of evidence support this prediction. According to the US Environmental Protection Agency, almost 28.9% of the country's greenhouse gas emissions in 2017 came from the transportation sector (Iclodean et al., 2017).

Specifically, light-duty vehicles were responsible for 60% of the emissions, while medium and heavy-duty trucks contributed 23% to the overall total. Europe additionally recorded a comparable figure during that year. The transportation sector's substantial contribution to greenhouse gas emissions in the 28 member states of the European Union exacerbated environmental concerns. Road transportation was the primary factor responsible for this contribution, with 72% of the total emissions.

- EVs primarily rely on batteries as their main energy source, although certain types of EVs may also utilize other energetic components. Therefore, some electric vehicles run solely on electricity, while others also have an internal combustion engine (ICE). Based on this, the International Electrotechnical Commission's (IEC) Technical Committee 69 has developed a classification system for battery-powered electric vehicles.
- Battery-electric vehicles (BEVs). Batteries are the only source of power that the drive train receives. An ICE system doesn't exist (Sanguesa et al., 2021).
- Hybrid electric vehicle (HEV). This relates to vehicles that utilize multiple energy sources or storage systems, with one of them being supplied with electrical energy. Due to this definition, there are numerous possible configurations for HEVs. One commonly used approach involves combining an internal combustion engine (ICE) with an electric power train that is powered by a battery. The operation of each system relies on the driving condition. Consequently, the hybrid electric vehicle (HEV) relies on the electrical component when traveling at low speeds, but changes to the internal combustion engine (ICE) for high-speed driving. In certain scenarios, both systems have the potential to work together and improve overall performance.
- Plug-in Hybrid Electric Vehicle (PHEV). This vehicle mostly depends on an electric powertrain for its operation. When the battery is completely drained, the vehicle seamlessly transitions to using the internal combustion engine (ICE). The primary goal of the ICE is to optimize the distance covered (Ding et al., 2017).

An electric vehicle (EV) can be directly connected to the electrical grid to charge its battery. The vehicle's batteries differ depending on the manufacturer. Table 3.1 provides an overview of the key characteristics of the batteries used in the most widely used electric vehicles (Hannan et al., 2014).

Table 3.1. Different EVs Characteristics of Battery

Model	**Vehicle Type**	**Energy Available (kWh)**	**Battery Size (kWh)**	**Energy Consumption (kWh/km)**	**Range (KM)**
Nissan Leaf	EV	37	39.5	0.165	240
Toyota RAV4	EV	32.18	41.8	0.20	160
Volkswagen Golf-e	EV	32	35.8	0.168	190
BMW i3	EV	27.2	33	0.13	200
Tesla model 3 standard range	EV	46	55	0.148	310
Tesla model S performance	EV	95	100	0.186	510
Audi e-tron	EV	83.6	95	0.232	360
Toyota prius	PHEV	7	8.8	0.175	40
BMW 530e	PHEV	8	9.2	0.237	34
Chevrolet volt	PHEV	13.7	17.1	0.21	64
Mitsubishi outlander	PHEV	11	13.8	0.29	37

Source: Triviño-Cabrera; Creative Commons License.

3.2. BATTERY TECHNOLOGY IN EVS

Learning Objectives:

- Understand the classification and characteristics of batteries used in electric vehicles.
- Understand the function and importance of Battery Management Systems (BMS) in EVs.

It is evident that electric vehicles, including BEVs, HEVs, and PHEVs, depend on batteries for their operation. Nevertheless, the capacities of batteries in BEVs and PHEVs vary, with BEVs having a greater capacity than PHEVs. Furthermore, the capacity of batteries in PHEVs surpasses that of batteries used in HEVs. As an example, the Tesla Model 3 Standard Range is an electric vehicle equipped with a 55-kWh battery (Liu et al., 2022). EVs primarily rely on batteries as their main storage component. These components enable the storage of energy in chemical form and its conversion into electrical energy as needed. The energy and power density of the batteries are determined by the technology in use. While numerous electric vehicles possess qualities that align with the requirements, power often poses limitations for certain functions like acceleration and regenerative braking. One way to address this limitation is by incorporating additional technologies like super capacitors (Miao et al., 2019).

The battery technology utilized in electric vehicles has undergone significant advancements, particularly in recent years due to the rise of prominent manufacturers in this industry. Figure 3.1 provides a visual representation of the battery classification employed in electric vehicles (EVs).

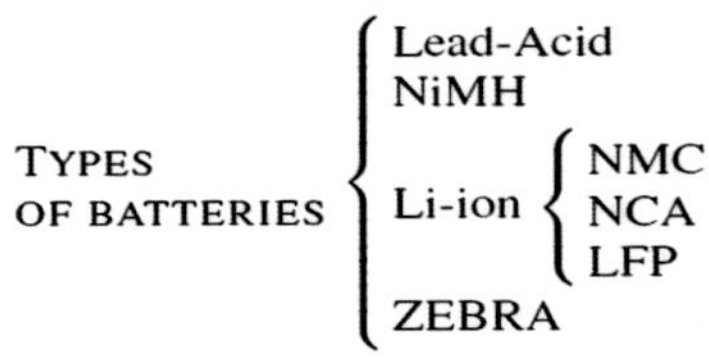

Figure 3.1. *Illustration of the types of batteries used in EVs.*

Source: José A. Aguado; Creative Commons License.

The first batteries used to start internal combustion engines in cars were lead-acid batteries. Many automakers have tested this technology in battery electric cars (BEVs), including General Motors and Toyota. Nevertheless, the limited energy density of this battery type makes them unsuitable for exclusive use in electric vehicles (Andwari et al., 2017).

ZEBRA batteries, commonly called molten salt batteries, have been utilized in numerous vehicle designs including urban bus models. The energy density of these batteries is significant, but they must be operated at high temperatures—between 270 and 350 ◦C. Due to this limitation, the use of this technology is only practical in vehicles that operate continuously to ensure the optimal working temperature.

The NiMH technology has gained significant popularity in various industries. Although this technology may have lower efficiency and slightly more weight compared to others, its impressive power density and energy, along with its simplicity, affordability, and long useful lifetime, make it an excellent solution for HEVs and PHEVs. Nevertheless, Li-ion batteries have emerged as the dominant choice in the market due to their exceptional electrical characteristics. Amongst this group of batteries, there is a diverse range of different types. Li-ion batteries come in a wide variety, and they are all made of different materials at the anode and cathode. Despite being widely used in consumer applications, lithium-cobalt oxide (LCO) is not thought to be a safe choice for electric vehicles (Manzetti & Mariasiu, 2015).

However, lithium-nickel-manganese-cobalt (NMC), lithium-nickel-cobalt-aluminum (NCA), and lithium-iron phosphate (LFP) are currently the most popular choices for automotive applications. Figure 3.2 presents a comprehensive analysis of the primary electrical characteristics of Li-ion batteries, including durability, energy density, power output, safety, and overall performance.

It is crucial to comprehend that each cycle of charging and discharging lithium-ion batteries leads to degradation and a decrease in capacity (Catenacci et al., 2013).

NMC batteries are commonly utilized in electric vehicle (EV) applications due to their ability to offer a high energy density. Their weight may thus be reduced. Additionally, they offer a degree of performance that is suitable in terms of cost, longevity, safety, and efficiency. Nissan Leaf, Chevrolet Bolt, and BMW i3 all use this particular kind of battery. In the past, NCA batteries have typically had a higher price tag compared to NMC batteries. Nevertheless, Panasonic and Tesla have collaborated to decrease the cobalt needed for this particular form of electrical storage. Consequently, the prices of these batteries have experienced a significant decrease, positioning them as the most affordable option currently available in the market (Sun et al., 2020).

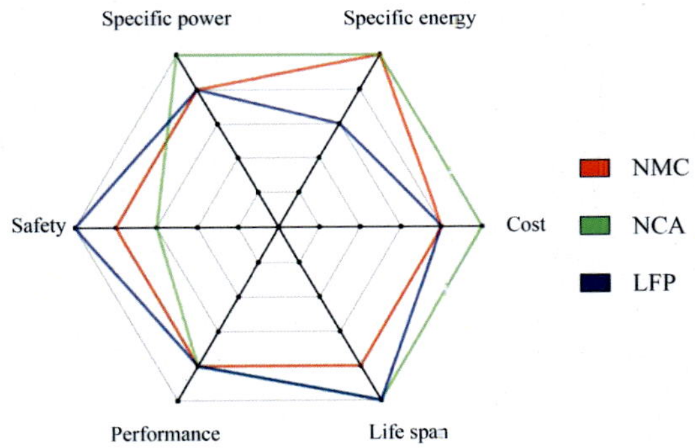

Figure 3.2. *Illustration of characteristics of Li-ion battery technologies.*

Source: José M. González; Creative Commons License.

On the other hand, LFP batteries are primarily utilized in larger vehicles such as buses, where the battery weight is of minimal concern. During these circumstances, the relatively low energy density of LFP batteries does not play a major role. However, they do provide a prolonged lifespan and operate safely. Table 3.2 provides a concise overview of the key electrical characteristics of the batteries that have been previously discussed (Neubauer et al., 2014).

When working on a charging system, it may not always be necessary or practical to create a highly complex battery model. However, there are various ways to model a vehicle battery. During a charging process, the prevailing model is typically represented by an equivalent resistance. The resistance value is determined by the battery's current, power, and voltage specifications (Song et al., 2020).

Table 3.2. Comparative Chart of Electric Vehicle Batteries

Type of Battery	Nominal Voltage (V)	Energy Density (Wh/kg)	Life Cycle	Specific Power (W/kg)	Self-Discharge (% per month)	Operating Temperature (◦)	Production Cost ($/kWh)
Lithium-ion (Li-ion)	3.6	118–250	2,000	200–430	<5	–20 to 60	150
Lead acid (Pb-acid)	2.0	35	1,000	180	<5	–15 to +50	60
ZEBRA	2.6	90–120	>1,200	155	<5	–245 to +350	230–345
Nickel-metal hydride (NiMH)	1.2	70–95	<3,000	200–300	20	–20 to +60	200–250
Lithium-iron phosphate (LFP)	3.2	120	>1,200	2,000–4,500	<5	–45 to 70	350

Source: Vahid Vahidinasab; Creative Commons License.

3.2.1. Battery Management System

Electric vehicles (EVs) extensively utilize lithium-ion batteries. This battery variant exhibits a reasonable level of energy density and little self-discharge, while also delivering a greater voltage per individual cell. Controlling a lithium-ion cell's charge and discharge process is essential, despite its many advantages. Excessive current, voltage, or charging can have detrimental effects on the battery's lifespan and pose safety risks, including fires or explosions (Shen & Gao, 2019).

Due to this reason, the vehicle is equipped with a Battery Management System (BMS) to ensure that the charging and discharging operation is conducted in compliance with safety rules. The BMS monitors temperature, voltage, output current, State of Health (SoH), input current, and State of Charge (SoC) among other battery characteristics in this specific role (Sadeghian et al., 2022).

By taking accurate measurements and estimates, the BMS can activate safety circuits when necessary to prevent excessive currents, overvoltages, overheating, or unexpected discharge. This proactive approach significantly extends the lifespan of the battery. Field Effect Transistors (FETs) enable the battery to be connected and disconnected in a controlled manner. Whenever a circumstance occurs that has the potential to damage the battery, cutoff FETs are engaged to fully separate the battery from other electrical circuits as a safety measure (Szumanowski & Chang, 2008).

The different battery components' current and voltage flows are controlled by the Battery Management System (BMS). Cells are the general term for these constituents. Energy can be conveniently stored in cells in an electrochemical configuration and turned into electrical energy via a reduction-oxidation process. Figure 3.3 illustrating a common BMS configuration where a single controller manages both the charging and discharging of the storage component. Certain approaches choose to utilize an individual controller for each process (Xing et al., 2011).

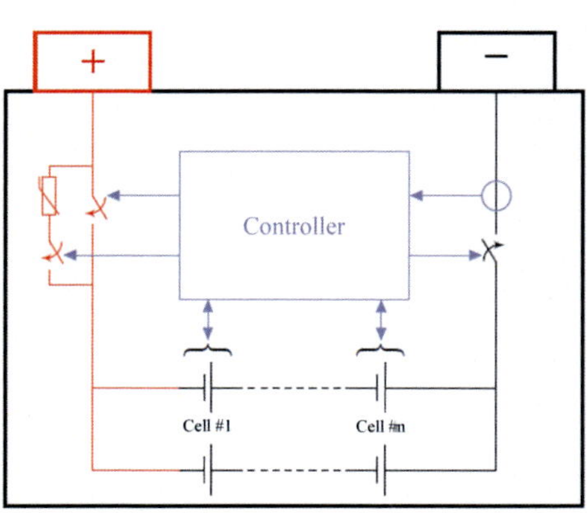

Figure 3.3. *Illustration of a BMS.*

Source: Zhongjing Ma; Creative Commons License.

Other than providing protection, the BMS also plays a crucial role in monitoring and balancing the cells. These functions are crucial as they help extend the battery lifespan by ensuring that the cell voltages remain similar. Monitoring systems are capable of quantifying the voltage in individual cells. By utilizing the gathered data, the Battery Management System (BMS) may make well-informed decisions and effectively accomplish cell balance. Two approaches can be used to achieve cell balancing. Passive or resistor-based balancing is a method that requires utilizing a dissipative device to discharge cells with excess voltage (Xiong & Tian, 2018) (Figures 3.3–3.5).

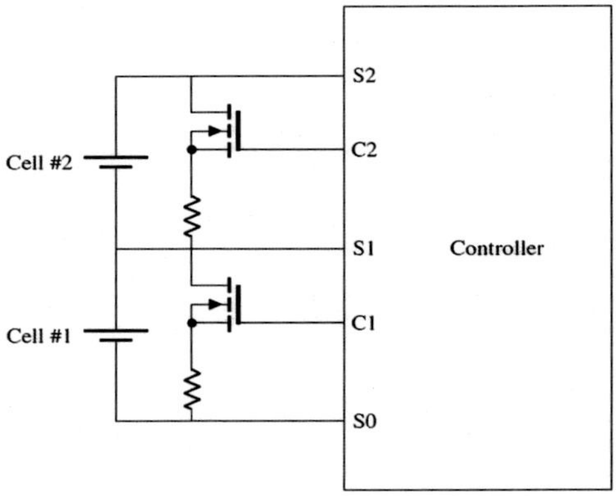

Figure 3.4. *Illustration of by-pass FETs in a passive balancing configuration.*

Source: Suli Zou, Creative Commons License.

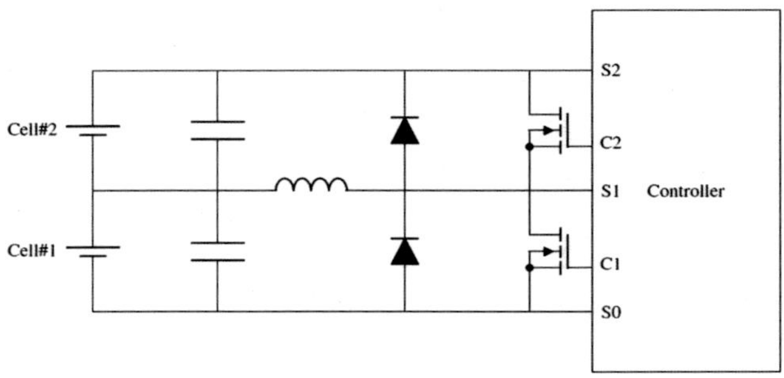

Figure 3.5. *Illustration of by-pass FETs in an active balancing configuration.*

Source: Vahid Vahidinasab; Creative Commons License.

To accomplish this objective, bypass FETs can be incorporated into the Battery Management System (BMS), as depicted in Figure 3.4. Due to its simplicity, it is frequently used in inexpensive solutions. The primary drawback of the passive technique is the energy dissipation that occurs during the cell balancing process, as well as the duration of time needed for this operation. An active technique is used to achieve a more intricate solution for cell balancing. By using this approach, the energy is evenly distributed across the cells to achieve energy equilibrium. This action can be executed during both the charging and discharging phases (Chatzakis et al., 2003). Electric vehicles (EVs) typically incorporate active cell balancing in their design (Figure 3.5).

Did You Know?

The first wireless electric vehicle charging system was installed in 2010 at the University of Maryland, displaying the potential for wire-free charging solutions.

Ultimately, the Battery Management System (BMS) is tasked with the transmission of battery status data to other interconnected electrical components. Possible communication interfaces include I2C, MODBUS, SPI, or IEEE 802.16. Based on this data, the charge/discharge process is modified to align with the real-time information, ensuring that the battery operates within its electrical limits and, consequently, extends its lifespan (Rahimi-Eichi et al., 2013).

3.3. CHARGING MODES IN EVS

Learning Objectives:

- Understand the classification of charging levels and modes in EVs.

The Battery Management System (BMS) regulates the division of energy among the individual cells of the battery, with a specific emphasis on its internal functions. The charging modes approach the battery as a cohesive entity, capable of receiving an internally distributed current and/or voltage through the BMS. The batteries can undergo charging using several methods. Within this specific context, the Society of Automotive Engineers (SAE) has established three charging levels in its Standard J1772:

1. **Level 1:** The device offers charging capabilities using a 120 V AC connection, which is compatible with a typical household electrical outlet. Therefore, no specific electrical equipment needs to be installed before in order to charge the car (Zhang et al., 2018).
2. **Level 2:** The charging process works by using a 240 V AC plug with a current of 40 A.

3. **Level 3:** As compared to earlier levels, the charging process in this case utilizes direct current (DC) rather than alternating current (AC). The voltage output is 480 volts.

The IEC standard 61851 offers an alternative categorization of the techniques used to charge an electric vehicle battery. In particular, the Commission identified four distinct modes:

1. **Mode 1:** The maximum voltage and maximum current that this gadget can handle is 250 V and 16 A, respectively, for AC charging. This imposes a limitation on the upper limit of power, capping it at 3.7 kilowatts. This connection can be supported by a standard household socket. It is advised to avoid using the device for long periods of time to prevent overheating.
2. **Mode 2:** Similar to mode 1, this is done using a standard plug, but the cord has a safety feature. The protection system's objectives are to track the charge/discharge amplitude and integrate a communication protocol. The maximum permissible current in this mode is 32 A. While the plug is ordinarily attached to the vehicle's Mennekes connection, it is connected to the mains via a Schuko connection (Liu & Zhou, 2022).
3. **Mode 3:** Unlike the previous modes, this one needs a specialized installation that is linked to the grid. This technology, which goes by the name "wall box," has both protection and control features. A dynamic communication channel between the car and the charging station is another feature of this mode. Installing control systems is the best equipment for smart grids because of its execution (Jiang et al., 2022).
4. **Mode 4:** This allows a direct current power transfer to take place, charging the car. The grid-connected wall box transforms the AC power into DC. The essential controls and safeguards are also included in this system. With a maximum power output of 240 kW and a high maximum current of 400 A, this application is meant to make quick charging easier. Owing to the wall box's distinctive features, its cost is quite expensive (Mao et al., 2019).

The levels and modes outlined earlier explain transferring power from the grid to the vehicle. In addition, there may be certain voltage and/or current requirements that need to be considered when charging the batteries. Considering this, charging can be done in one of four ways:

1. **Constant-Current (CC) Charging:** Throughout the charging procedure, the battery receives a steady current. Usually, the steady

current is maintained at a low level to increase the battery's longevity. One downside to this is that the charging process can be quite time-consuming. Typically, the charging process comes to an end once a predetermined time has passed. This duration is determined by factors such as the current being used and the capacity of the battery. This approach can be followed by NiMH and Li-ion batteries.

2. **Constant-Voltage (CV) Charging:** A fixed voltage is applied in order to charge the batteries. One major benefit is the prevention of over-voltages and the associated damages. This omission is expected to extend the lifespan of the battery. The current is progressively decreased in this charging mode until it reaches a preset value. Under these conditions, the charging process is considered complete (Shareef et al., 2016).
3. **Constant-Current–Constant-Voltage (CC–CV) Charging:** There are two phases to the charge generation. A constant current is used in first phase. The parameter, which indicates the maximum current that the battery can withstand without experiencing any harm, is calculated in accordance with the manufacturer's instructions. At this time, there is an increase in voltage. After the voltage reaches its peak, the charging mode smoothly transitions into a constant voltage mode. During the second phase, the current experiences a gradual decrease. It is widely accepted that the charging process is considered complete once the current reaches a predetermined level. Shown in Figure 3.6 is a commonly used charging mode for EV batteries (Bhatti et al., 2016).
4. **Multi-Stage Constant-Current (MCC) Charging:** In this method, the battery is charged in stages. Every stage requires the implementation of a consistent flow of electric current. The value is determined by the phase. Once the voltage reaches a specific threshold, the phase transitions to a different current. The phase is a critical factor in determining this parameter. One of the primary advantages of this method is its capacity to deliver fast charging. That is why it is primarily utilized in fast chargers. Nevertheless, achieving an optimal configuration for the parameters of each phase (threshold and current voltage) can be quite challenging (Aretxabaleta et al., 2021).

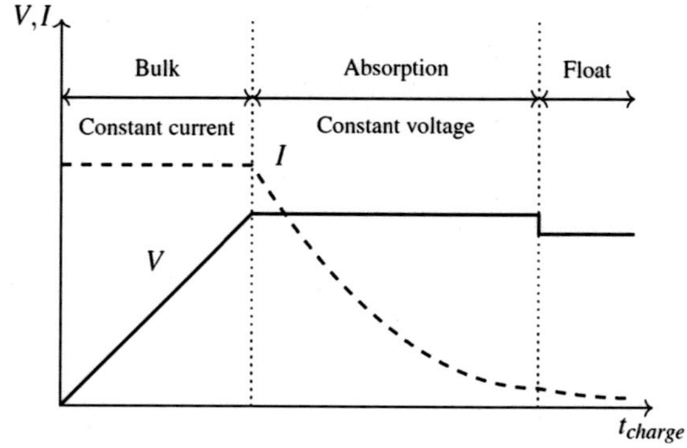

Figure 3.6. *Charging process of battery in CC–CV mode.*

Source: Chong Zhu; Creative Commons License.

Table 3.3 presents a comparison of the charging approaches.

Table 3.3. *Charging Mode Approaches*

Mode	Key Elements	Advantages	Disadvantages
CV	Charging constant voltage; terminal condition.	Simple to execute; table terminal voltage.	Simple to trigger the battery's lattice collapse.
CC	Charging constant current rate; terminal condition	Easy to implement	Capacity utilization is low
CC–CV	Constant current rate in CC phase; constant voltage in CV phase; terminal condition	Capacity utilization is high; stable terminal voltage	Challenging goals to balance, like energy loss, charging speed, and temperature change.
MCC	The number of CC stages; constant current rates across every phase.	Implementation is easy; fast charging is easily achievable.	Challenging to balance objectives such as capacity utilization, battery lifetime, and charging speed.

Source: Chong Zhu; Creative Commons License.

3.4. WPT OPERATION MODES

Learning Objectives:

- Understand the operation modes of Wireless Power Transfer (WPT) for EVs.

Wireless chargers provide greater flexibility in charging EVs in a wider range of scenarios, including parked, temporarily halted, and even moving. As a result, by reducing the battery's capacity, the price and weight of the battery can be decreased (Jiang et al., 2018).

Wireless charging can be categorized into three modes: static, stationary, and dynamic. Figure 3.7 presents a graphical depiction of the three operation modes.

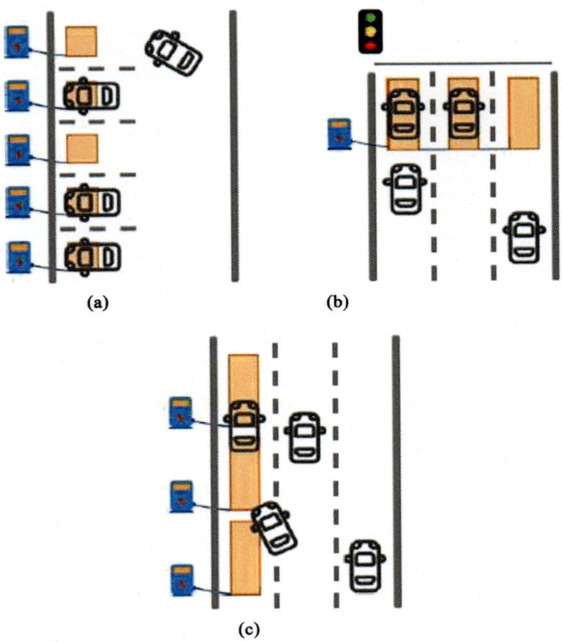

Figure 3.7. *Illustration of WPT operation modes – (a) static; (b) quasi-dynamic; and (c) dynamic.*

Source: Noaki Shenora; Creative Commons License.

Static wireless power transfer occurs when the charging process is limited to a fixed location and requires the vehicle to be powered off during the entire charging period. This relates to chargers that are situated at homes or car parks. Advanced features that offer advice on how to position the car optimally to prevent coil misalignment may be included in the chargers. In Berlin, Bombardier has successfully developed a magnetic-resonance bus charger that can generate over 200 kW of electricity when in a stationary operating mode (Fu et al., 2021).

The quasi-dynamic or stationary charge is another kind of WPT charge. These two particularities apply to this kind of fee. First of all, the engine is still running even though the car is not moving. Second, this state of affairs persists for a little duration that is insufficient to achieve a complete charge. This type of charge helps to replenish the energy in the vehicles used for public transportation when stopping at bus/tram stations or taxi ranks (Zhao et al., 2016).

This was the situation with the 2013 Mannheim, Germany, route-63 bus trial. The bus's batteries could be charged while it picked up people, so it didn't have to stop and charge them. To make this operation possible, the bus lane was changed appropriately. Moreover, private automobiles could be recharged while stopped by traffic signals or regular traffic patterns (Limb et al., 2018).

The dynamic mode, which describes the charging that occurs while the car is moving, i.e., when it is typically traveling on a paved road, is an important operation mode in WPT for EVs. In particular, electronics would be installed in some sections of specific lanes to allow WPT for usage by private automobiles and by buses and trams used for public transportation. This kind of charging encourages the use of RPEVs, or highway-powered electric vehicles.

Dynamic charging has been successfully implemented in several cities, including those in South Korea and Spain, through innovative projects such as the OLEV and Victoria projects. A prototype that charges a bus at stops and at the conclusion of its trip using magnetic resonance technology has been successfully installed by Conductix-Wampfler in Torino, Italy (Chen et al., 2017).

Numerous initiatives have been put into action to implement wireless technologies for EV charging. Here are two examples that can be used as references. FABRIC conducted a comprehensive study on the widespread integration of electric vehicles in upcoming transportation networks. For electric vehicles to become widely used, they need to have well-developed technology and advanced charging choices that make driving them like driving a regular car. FABRIC's primary focus was addressing the issue of range anxiety, which arises from the limited range that current electric vehicles experience.

Electric vehicles have the potential to gather energy from the road in the future, either through a conductive or contactless method known as dynamic charging. In contrast to the existing models with bigger storage capacity, rapid charging or interchangeable batteries, innovative on-road charging solutions will enhance the driving range and longevity of electric vehicles, while also improving their energy efficiency and cost-effectiveness, as they require a smaller battery (Yang et al., 2019).

Another noteworthy project is the UNPLUGGED project. This research examined the impact of inductive charging on electric vehicles in urban settings, focusing particularly on enhancing the convenience and eco-friendliness of car-based transportation. In order to improve customer acceptance and perceived usefulness, the study concentrated on analyzing how well smart inductive charging infrastructure integrates electric vehicles into metropolitan road systems. Inductive charging's technical viability, interoperability, user perception, practical factors, and socioeconomic consequences were all thoroughly evaluated by UNPLUGGED.

3.5. VEHICLE TO GRID TECHNOLOGY

Learning Objectives:

- Understand the concept and implications of Vehicle-to-Grid (V2G) technology.

The advantages of utilizing electric vehicles extend beyond just sustainability, as this technology will also have a significant impact on smart grids. From this perspective, electric vehicles can be viewed as portable power sources with minimal inertia. When necessary, electric vehicles can quickly supply energy to the electrical grid. This function makes managing electrical grids easier, particularly in grids that have a large amount of renewable energy sources. When there is a surplus of renewable electricity being generated in the grid, it can be used to recharge vehicle batteries. Moreover, the battery has the capacity to offer supplementary functions such as voltage or frequency adjustment (Donadee et al., 2019).

Therefore, the battery of an electric vehicle could serve a dual purpose: either being charged or functioning as a generator. These operations are commonly referred to as V2G operations. A number of grid-related and battery-related issues are important in assessing whether V2G operations are feasible. State-of-the-art algorithms consider these parameters and offer suggestions on the optimal timing and location for carrying out the charge/discharge operations (Habib et al., 2015).

Did You Know?

Wireless Power Transfer (WPT) systems for electric vehicles operate on the principle of magnetic resonance or inductive charging, where electricity is transferred between coils via magnetic fields.

In 2017, the V2G capacity reached 2.6 GW, as reported by the research agency Navigates. This estimate only represents a small portion of the larger global grid services market. However, the figure is expected to increase and is estimated to reach a power capacity of 20.5 GW (4% of the grid services) by 2026. Wireless power transfer technology presents the exciting potential for enabling autonomous charging and discharging of electric vehicles. This feature is particularly useful for V2G operations, allowing the charge/discharge process to be initiated during periods when the user cannot actively participate (Alsharif et al., 2021).

In order to advance these operations, it is necessary to design and implement wireless chargers that can facilitate power flow in both directions: from the grid to the battery and vice versa. Power converters used in bi-directional wireless power transfer (BWPT) are more complex compared to those used in one-way wireless chargers (Uddin et al., 2018).

3.6. WIRELESS POWER TRANSFER MARKET PERSPECTIVES

Learning Objectives:

- Understand market dynamics and key players in Wireless Power Transfer.
- Know about leading technologies and partnerships in wireless charging.

Recent market forecasts suggest a positive outlook for wireless EV charging markets, particularly in transportation applications. Several factors, including consumer preferences for hassle-free charging, the requirement for charging infrastructure, and the desire for efficient and reliable charging options, are expected to increase the demand for wireless EV charging (Bi et al., 2016).

The Asia Pacific area is predicted to overtake North America as the market leader for wireless EV charging systems by 2025, with the former leading the way in the market for wireless transfer applications. Anticipated growth in the European Union is projected to be substantial between 2020 and 2024.

The number of wireless EV chargers on the market has significantly increased, and a variety of solutions are currently provided by start-ups, well-established technology companies, and automakers. The dynamic charging sector is projected to experience rapid growth in the wireless EV charging market from 2020 to 2025 (Feng et al., 2020).

Wireless EV technology vendors and EV manufacturers are forming partnerships to investigate business strategies, evaluate market niche viability, and test commercial reliability. Various providers are engaged in a competitive race to introduce a range of initiatives, from development and prototype testing to actual deployment, showcasing different levels of technological advancement.

Next, an overview of various technology providers and car manufacturers' solutions for wireless EV chargers will be provided (Miller et al., 2014).

3.6.1. WiTricity

In 2007, WiTricity was founded as an MIT start-up. It uses magnetic resonance technology to develop wireless power solutions. One of the top technology suppliers for wireless applications is WiTricity. It collaborates with suppliers and manufacturers of EV vehicles to implement EV wireless charging.

After acquiring the Qualcomm Halo IP portfolio, WiTricity has solidified its position as a leading provider of wireless charging technology to car manufacturers (Gozalvez, 2007).

For WiTricity's EV charging products, the overall system efficiency of 90–93% is quite good. WiTricity collaborates closely with worldwide standards-setting organizations like IEC/ISO and SAE worldwide. WiTricity EV wireless products provide multiple charging speeds to accommodate various EV battery types. With a unified system design, WiTricity products can conveniently charge vehicles of different ground clearance levels, from low to high (Dai et al., 2017).

Witricity is affiliated with the AirFuel Alliance. This group is made up of prominent companies dedicated to promoting the adoption of wireless power technology. This organization is dedicated to enhancing the user experience by developing standards for wireless charging, specifically focusing on magnetic resonance technology. The aim is to create a comprehensive environment that fosters the development of inter-operability and open standards, which in turn promotes innovation and advances the marketplace (Kumar et al., 2015).

3.6.2. Qualcomm

In 2011, Qualcomm acquired the "Halo" technology from the University of Auckland. The dynamic test track relies heavily on this component for the purpose of charging electric vehicles. Qualcomm Halo created a device that uses magnetic resonance to connect the electric grid to the EV pad, which is an essential component of the car charging system. Modern car-to-grid technology enables the Qualcomm Halo system to move energy from the battery of an electric vehicle to the power grid (Lakoff, 2008). This technology has an exceptional overall efficiency of over 90% and a power range of 3.3 to 20 kW. Even when the charging pads are not exactly aligned, Qualcomm's technology uses special techniques to transfer power very well. Witricity purchased Qualcomm Inc.'s Halo wireless charging division in February 2019. This agreement covers a wide range of wireless charging technology-related patents and applications. Qualcomm Halo technology will be integrated by WiTricity into its current wireless charging operations (Prabhu, 2019).

3.6.3. EVATRAN

As a wireless EV charging option, Evatran's Plugless Level 2 EV Charging System has been successfully marketed. Evatran initiated the Plugless field trials in 2011 and implemented the technology in various locations throughout the United States, including pilot programs with Hertz, SAP, and Google. Since the successful delivery of its first production system in the beginning of 2014, Plugless devices have conveniently charged for more than a million hours (Carlson & Normann, 2014).

The collaboration has provided services to the US Department of Energy and the Swedish Energy Agency as part of its agreement with Oak Ridge National Laboratory. In 2016, VIE & Evatran established a collaboration to address the requirements of original equipment manufacturers (OEMs) manufacturing electric vehicles (EVs) for the Chinese market (Sasikala & Ranjan, 2023).

In December 2016, the company showcased a functional 7.2 kW wireless charging system that successfully supplied electricity to a BMW i3 car with a ground clearance of 254 mm.

3.6.4. Eaton HEVO

HEVO's mission is to establish a universal standard for wireless charging of electric vehicles, catering to the needs and expectations of users who desire a convenient and effortless charging experience. HEVO provides a secure and efficient solution for charging electric vehicles by offering a wireless charging option. By using this method, plug-in charging's usual hazards and inconveniences are eliminated (Patil et al., 2017).

Eaton Corporation announced in December 2013 that a scalable HyperCharger with a 200 kW to 1 MW power range would be available for fast charging hybrid and electric buses and vehicles. According to various press articles, installations have reportedly taken place in Florida, Tallahassee, Massachusetts, Worcester, and Stockton, California, specifically for the purpose of utilizing Proterras EcoRide bus. It seems that the electric bus range has been extended to 240 miles per day with the help of 8 en-route charges (Lassioui et al., 2021).

3.6.5. Momentum Dynamics

For the transportation and automotive industries, Momentum Dynamics develops high-power wireless power transfer solutions that are efficient. The scientific foundation of the technology is magnetic resonance WPT. With the Momentum Charger, electric vehicles of all kinds may be charged in all weather and without monitoring (Denny et al., 1998). Momentum's inductive charging technology has demonstrated efficacy at power levels above 200 kW for in-transit bus applications. The system can deliver up to 450 kW. Systems that could transmit 30 kW of power across a 12-inch air gap in rain or snow are undergoing field trials.

3.7. EV WIRELESS CHARGER GENERIC STRUCTURE

Learning Objectives:

- Understand the structure and operation principles of EV wireless chargers.

Magnetic-resonance technology is the most commonly used technology for wireless chargers in electric vehicles (EVs). The diagram of a typical charger of this type is shown in Figure 3.8.

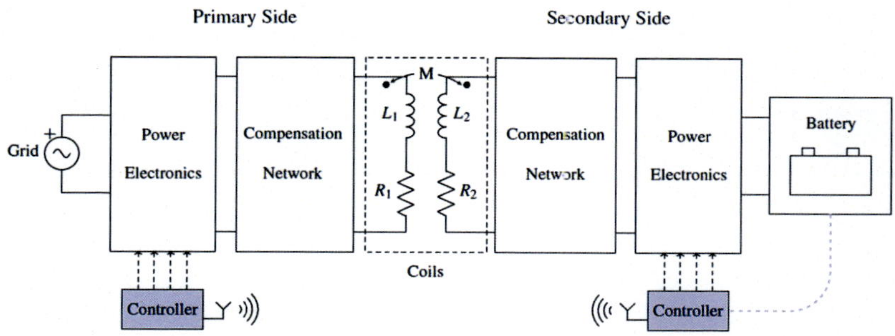

Figure 3.8. *Illustration of generic structure of an EV wireless charger.*

Source: Robert Shorten; Creative Commons License.

The primary and secondary side are the two main parts of the wireless charger. The main part comprises the parts connecting the pavement-embedded coil to the utility grid. The secondary side is responsible for connecting to the components that are installed in the vehicle (Kan et al., 2018).

This type of wireless charger is comprised of two pairs of coils. The primary coil is commonly known as the transmitter coil, whereas the secondary coil is also referred to as the receiver or pick-up coil. The primary coil is stimulated with a current that varies over time, using magnetic resonance technology. Consequently, the generation of a magnetic field occurs. When the magnetic field passes through the secondary coil, it generates a voltage in its terminals. The efficiency of voltage induction is greatly affected by the geometry and composition of the coils, as they play a crucial role in utilizing the magnetic field.

Practice Problem

A company is planning to implement wireless charging stations for its fleet of electric vehicles. The vehicles in the fleet have different battery capacities ranging from 40 kWh to 100 kWh. If the charging efficiency of the wireless charging system is 90%, how much energy (in kWh) will be required to fully charge a vehicle with a 60-kWh battery capacity?

Solutions to Practice Problem

Battery capacity of the vehicle = 60 kWh
Charging efficiency of the wireless charging system = 90%
Energy required for full charge = Battery capacity/Charging efficiency
Energy required = 60 kWh/0.90 = 66.67 kWh
Therefore, approximately 66.67 kWh of energy will be required to fully charge a vehicle with a 60-kWh battery capacity.

The system functions based on resonance to optimize the transfer of power. The coils are connected to two compensating networks on either side in order to do this. The compensation network's ability to tolerate coil misalignment and operating frequency fluctuations is largely dependent on its design and layout. The key compensation networks are covered in detail in Chapter 4.

Because the utility grid's frequency is too low to ensure a suitable rate of variation in magnetic flux, power converters are necessary (Chen et al., 2019).

The secondary power converter transforms the AC signal sent between the coils during wireless charging into a DC signal. The car's battery is then charged using this DC signal. On the other hand, the primary power converter increases the frequency of the grid. This wireless power transfer typically uses a magnetic field that runs at frequencies in the tens of kilohertz range.

As mentioned earlier, the relevant international organizations are working on standards that suggest setting this parameter at 85 kHz. In addition, power electronics are valuable for adapting the power, current, and voltage to meet the battery's needs. For optimal operation of the wireless charger, it is essential to continuously monitor certain electrical parameters, such as the battery status provided by the BMS. After processing the data, the controllers need to make adjustments to the configuration of certain devices. For example, a controller can manipulate timing of the transistors in the power converters (Panchal et al., 2018).

Did You Know?

Wireless charging technology is not only convenient for individual electric vehicle owners but also holds potential benefits for autonomous electric vehicles, allowing for automated charging processes without human intervention.

Examine the coils and compensation networks on the power converters, making them simpler. Several analogous circuits are obtained with this simplification (Bilal et al., 2021). An AC voltage source takes the place of the power converters on the primary side. This source oscillates at a magnitude and rate that coincide with the fundamental frequency of the output signal from the converter (Ramezani et al., 2018).

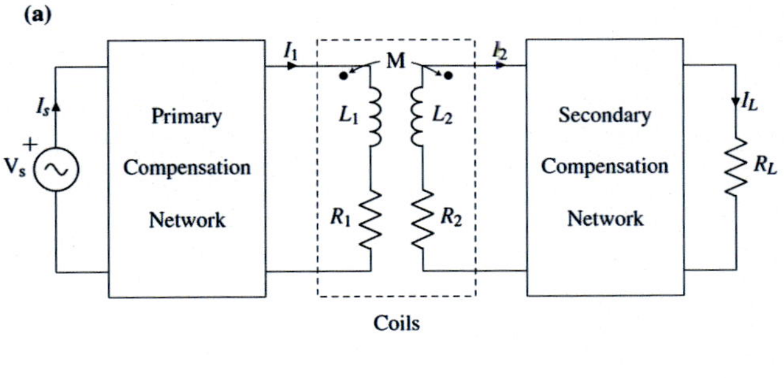

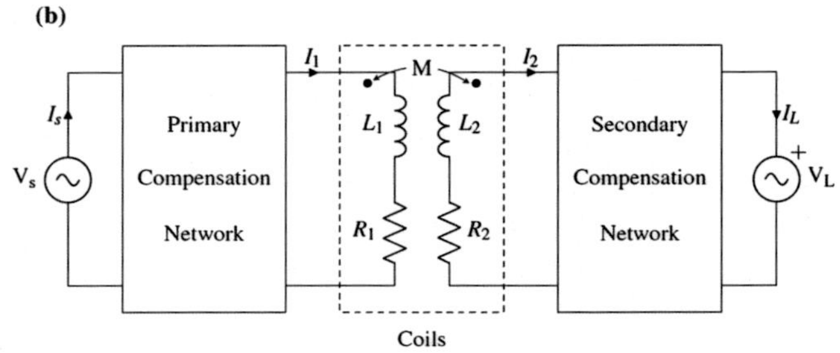

Figure 3.9. *Equivalent circuits for (a) charging mode in both uni-directional and bi-directional chargers; and (b) the bi-directional charger's discharge mode.*

Whether the charger is unidirectional or bidirectional determines how simple the secondary side power converters are to make. Load resistances are used to model the battery and secondary power converter in a unidirectional charger. This resistance's value is influenced by the battery's needed voltage and charging power. As an alternative, the battery is typically modeled as a DC voltage source in bi-directional chargers (Mahesh et al., 2021). The analogous circuit for both kinds of chargers is shown in Figure 3.9.

SUMMARY

- This chapter titled "Wireless Chargers for Electric Vehicles" provides a comprehensive overview of the different aspects related to wireless charging technology for electric vehicles (EVs). The discussion starts by exploring different types of electric vehicles, highlighting the importance of battery technology and battery management systems in the context of EVs.
- It delves into the examination of charging modes in EVs, and subsequently explores the advantages of wireless power transfer (WPT). The chapter provides detailed explanations of various operation modes in wireless power transfer and explores the concept of vehicle-to-grid technology.
- The compatibility and safety of EV wireless charging standards, such as SAE J2954 and IEC 61980/19363, are emphasized. An analysis of market views on wireless power transfer is conducted, highlighting notable companies such as Qualcomm, WiTricity, Eaton HEVO, and Momentum Dynamics. The efforts made by automakers to implement wireless charging technologies are also covered.
- A summary of the general design of EV wireless chargers wraps off the chapter, giving readers a thorough grasp of the state of technology related to wireless charging for electric cars.

REVIEW QUESTIONS

1. What are the main types of electric vehicles and how do their powertrain technologies differ?
2. How do battery technology and battery management systems impact the performance and efficiency of EVs?
3. What are the different EV charging modes and their benefits and limitations?
4. What benefits does wireless power transfer (WPT) offer for EVs compared to conventional charging?
5. What are the different operation modes of WPT systems and how do they suit various charging scenarios?
6. How does vehicle-to-grid (V2G) technology work and what are its potential benefits and challenges?
7. What key requirements are outlined in standards like SAE J2954 and IEC 61980/19363 for EV wireless chargers?

8. How are companies like WiTricity, Qualcomm, and Momentum Dynamics advancing EV wireless charging technology?
9. What are the essential components of EV wireless chargers and how do they work together?
10. What are the implications of widespread adoption of wireless chargers for EVs on the automotive industry, energy infrastructure, and environmental sustainability?

MULTIPLE CHOICE QUESTIONS

1. **What is one of the key components emphasized in electric vehicles (EVs) regarding battery technology?**
   a. Fuel cell efficiency
   b. Battery management system
   c. Transmission fluid quality
   d. Tire pressure monitoring
2. **What is the primary benefit associated with wireless power transfer (WPT) in electric vehicles?**
   a. Reduced charging speed
   b. Increased emissions
   c. Enhanced convenience
   d. Higher operating costs
3. **Which standard ensures interoperability and safety in EV wireless chargers?**
   a. SAE J2954
   b. ISO 9001
   c. ASTM D1186
   d. ANSI/TIA-568
4. **What technology is utilized for wireless power transfer (WPT) in electric vehicles?**
   a. Inductive charging
   b. Hydrogen fuel cells
   c. Solar panels
   d. Diesel generators

5. **What does EVATRAN specialize in within the realm of electric vehicle technology?**
   a. Battery recycling
   b. Autonomous driving systems
   c. Wireless charging solutions
   d. Aerodynamic design
6. **Which organization is responsible for the IEC standards related to EV wireless chargers?**
   a. International Energy Agency (IEA)
   b. International Electro Technical Commission (IEC)
   c. Institute of Electrical and Electronics Engineers (IEEE)
   d. International Standards Organization (ISO)

Answers to Multiple Questions

1. (b); 2. (c); 3. (a); 4. (a); 5. (c); 6. (b)

REFERENCES

1. Ahmad, A., Alam, M. S., & Chabaan, R. (2017). A comprehensive review of wireless charging technologies for electric vehicles. *IEEE Transactions on Transportation Electrification, 4*(1), 38–63.
2. Alsharif, A., Tan, C. W., Ayop, R., Dobi, A., & Lau, K. Y. (2021). A comprehensive review of energy management strategy in Vehicle-to-Grid technology integrated with renewable energy sources. *Sustainable Energy Technologies and Assessments, 47*, 101439.
3. Amjad, M., Farooq-i-Azam, M., Ni, Q., Dong, M., & Ansari, E. A. (2022). Wireless charging systems for electric vehicles. *Renewable and Sustainable Energy Reviews, 167*, 112730.
4. Andrei, M. (2020). The Romanian wireless power transfer network. *Technical Sciences, 5*(2), 149–156.
5. Andwari, A. M., Pesiridis, A., Rajoo, S., Martinez-Botas, R., & Esfahanian, V. (2017). A review of battery electric vehicle technology and readiness levels. *Renewable and Sustainable Energy Reviews, 78*, 414–430.
6. Aretxabaleta, I., De Alegria, I. M., Andreu, J., Kortabarria, I., & Robles, E. (2021). High-voltage stations for electric vehicle fast-charging: Trends, standards, charging modes and comparison of unity power-factor rectifiers. *IEEE Access, 9*, 102177–102194.

7. Bhatti, A. R., Salam, Z., Aziz, M. J. B. A., Yee, K. P., & Ashique, R. H. (2016). Electric vehicles charging using photovoltaic: Status and technological review. *Renewable and Sustainable Energy Reviews, 54*, 34–47.

8. Bi, Z., Kan, T., Mi, C. C., Zhang, Y., Zhao, Z., & Keoleian, G. A. (2016). A review of wireless power transfer for electric vehicles. *Applied Energy, 179*, 413–425.

9. Bilal, A., Kim, S., Lin, F., & Covic, G. (2021). Analysis of IPT intermediate coupler system for vehicle charging over large air gaps. *IEEE Journal of Emerging and Selected Topics in Industrial Electronics, 3*(4), 1149–1158.

10. Campi, T., Cruciani, S., Maradei, F., & Feliziani, M. (2019). Magnetic field during wireless charging in an electric vehicle according to standard SAE J2954. *Energies, 12*(9), 1795.

11. Carlson, R. W., & Normann, B. (2014). Test results of the PLUGLESS™ inductive charging system from Evatran Group, Inc. *SAE International Journal of Alternative Powertrains, 3*(1), 64–71.

12. Catenacci, M., Verdolini, E., Bosetti, V., & Fiorese, G. (2013). Going electric: Expert survey on the future of battery technologies for electric vehicles. *Energy Policy, 61*, 403–413.

13. Chatzakis, J., Kalaitzakis, K., Voulgaris, N. C., & Manias, S. N. (2003). Designing a new generalized battery management system. *IEEE Transactions on Industrial Electronics, 50*(5), 990–999.

14. Chen, Y., Kou, Z., Zhang, Y., He, Z., Mai, R., & Cao, G. (2017). Hybrid topology with configurable charge current and charge voltage output-based WPT charger for massive electric bicycles. *IEEE Journal of Emerging and Selected Topics in Power Electronics, 6*(3), 1581–1594.

15. Chen, Y., Zhang, H., Park, S. J., & Kim, D. H. (2019). A switching hybrid LCC-S compensation topology for constant current/voltage EV wireless charging. *IEEE Access, 7*, 133924–133935.

16. Cirimele, V., Torchio, R., Villa, J. L., Freschi, F., Alotto, P., Codecasa, L., & Di Rienzo, L. (2020). Uncertainty quantification for SAE J2954 compliant static wireless charge components. *IEEE Access, 8*, 171489–171501.

17. Dai, Z., Wang, J., Long, M., & Huang, H. (2017). A witricity-based high-power device for wireless charging of electric vehicles. *Energies, 10*(3), 323.

18. Denny, M., Gaylord, B., Helmuth, B., & Daniel, T. (1998). The menace of momentum: Dynamic forces on flexible organisms. *Limnology and Oceanography, 43*(5), 955–968.

19. Dimitriadou, K., Rigogiannis, N., Fountoukidis, S., Kotarela, F., Kyritsis, A., & Papanikolaou, N. (2023). Current trends in electric vehicle charging infrastructure; opportunities and challenges in wireless charging integration. *Energies, 16*(4), 2057.

20. Ding, N., Prasad, K., & Lie, T. T. (2017). The electric vehicle: A review. *International Journal of Electric and Hybrid Vehicles, 9*(1), 49–66.

21. Donadee, J., Shaw, R., Garnett, O., Cutter, E., & Min, L. (2019). Potential benefits of vehicle-to-grid technology in California. *IEEE Electrification Magazine, 7*(2), 40–45.

22. Feng, H., Tavakoli, R., Onar, O. C., & Pantic, Z. (2020). Advances in high-power wireless charging systems: Overview and design considerations. *IEEE Transactions on Transportation Electrification, 6*(3), 886–919.

23. Fisher, T. M., Farley, K. B., Gao, Y., Bai, H., & Tse, Z. T. H. (2014). Electric vehicle wireless charging technology: A state-of-the-art review of magnetic coupling systems. *Wireless Power Transfer, 1*(2), 87–96.

24. Fu, N., Deng, J., Wang, Z., Wang, W., & Wang, S. (2021). A hybrid mode control strategy for LCC–LCC-compensated WPT system with wide ZVS operation. *IEEE Transactions on Power Electronics, 37*(2), 2449–2460.

25. Gozalvez, J. (2007). Witricity—The wireless power transfer [mobile radio]. *IEEE Vehicular Technology Magazine, 2*(2), 38–44.

26. Habib, S., Kamran, M., & Rashid, U. (2015). Impact analysis of vehicle-to-grid technology and charging strategies of electric vehicles on distribution networks: A review. *Journal of Power Sources, 277*, 205–214.

27. Hannan, M. A., Azidin, F. A., & Mohamed, A. (2014). Hybrid electric vehicles and their challenges.

28. Hutchinson, L., Waterson, B., Anvari, B., & Naberezhnykh, D. (2019). Potential of wireless power transfer for dynamic charging of electric vehicles. *IET Intelligent Transport Systems, 13*(1), 3–12.

29. Iclodean, C., Varga, B., Burnete, N., Cimerdean, D., & Jurchiş, B. (2017). Comparison of different battery types for electric vehicles. In *IOP Conference Series: Materials Science and Engineering* (Vol. 252, No. 1, pp. 012058).

30. Jayalath, S., & Khan, A. (2020). Design, challenges, and trends of inductive power transfer couplers for electric vehicles: A review. *IEEE Journal of Emerging and Selected Topics in Power Electronics, 9*(5), 6196–6218.

31. Jiang, X., Zhao, L., Cheng, Y., Wei, S., & Jin, Y. (2022). Optimal configuration of electric vehicles for charging stations under the fast power supplement mode. *Journal of Energy Storage, 45*, 103677.

32. Jiang, Y., Wang, L., Wang, Y., Liu, J., Wu, M., & Ning, G. (2018). Analysis, design, and implementation of WPT system for EV's battery charging based on optimal operation frequency range. *IEEE Transactions on Power Electronics, 34*(7), 6890–6905.
33. Kan, T., Lu, F., Nguyen, T. D., Mercier, P. P., & Mi, C. C. (2018). Integrated coil design for EV wireless charging systems using LCC compensation topology. *IEEE Transactions on Power Electronics, 33*(11), 9231–9241.
34. Kim, D., Kim, H., Huang, A., He, Q., Zhang, H., Ahn, S., & Fan, J. (2019). Analysis and introduction of effective permeability with additional air-gaps on wireless power transfer coils for electric vehicle based on SAE J2954 recommended practice. *Energies, 12*(24), 4797.
35. Kim, J., Clerckx, B., & Mitcheson, P. D. (2020). Signal and system design for wireless power transfer: Prototype, experiment and validation. *IEEE Transactions on Wireless Communications, 19*(11), 7453–7469.
36. Kumar, A., Luthra, S. K., & Ratan, R. (2015). Witricity: Designed to change the world. *International Journal of Innovative Research in Electronics and Communication, 2*(3), 1–8.
37. Lakoff, S. (2008). Upstart startup: "Constructed advantage" and the example of Qualcomm. *Technovation, 28*(12), 831–837.
38. Lassioui, A., El Fadil, H., Rachid, A., Bouanou, T., & Giri, F. (2021). Adaptive output feedback nonlinear control of a wireless power transfer charger for battery electric vehicle. *Journal of Control, Automation and Electrical Systems, 32*, 492–506.
39. Li, S., & Mi, C. C. (2014). Wireless power transfer for electric vehicle applications. *IEEE Journal of Emerging and Selected Topics in Power Electronics, 3*(1), 4–17.
40. Limb, B. J., Asher, Z. D., Bradley, T. H., Sproul, E., Trinko, D. A., Crabb, B., & Quinn, J. C. (2018). Economic viability and environmental impact of in-motion wireless power transfer. *IEEE Transactions on Transportation Electrification, 5*(1), 135–146.
41. Liu, L., & Zhou, K. (2022). Electric vehicle charging scheduling considering urgent demand under different charging modes. *Energy, 249*, 123714.
42. Liu, W., Placke, T., & Chau, K. T. (2022). Overview of batteries and battery management for electric vehicles. *Energy Reports, 8*, 4058–4084.
43. Mahesh, A., Chokkalingam, B., & Mihet-Popa, L. (2021). Inductive wireless power transfer charging for electric vehicles—A review. *IEEE Access, 9*, 137667–137713.

44. Manzetti, S., & Mariasiu, F. (2015). Electric vehicle battery technologies: From present state to future systems. *Renewable and Sustainable Energy Reviews, 51*, 1004–1012.

45. Mao, T., Zhang, X., & Zhou, B. (2019). Intelligent energy management algorithms for EV-charging scheduling with consideration of multiple EV charging modes. *Energies, 12*(2), 265.

46. Mayordomo, I., Dräger, T., Spies, P., Bernhard, J., & Pflaum, A. (2013). An overview of technical challenges and advances of inductive wireless power transmission. *Proceedings of the IEEE, 101*(6), 1302–1311.

47. Miao, Y., Hynan, P., Von Jouanne, A., & Yokochi, A. (2019). Current Li-ion battery technologies in electric vehicles and opportunities for advancements. *Energies, 12*(6), 1074.

48. Miller, J. M., Onar, O. C., & Chinthavali, M. (2014). Primary-side power flow control of wireless power transfer for electric vehicle charging. *IEEE Journal of Emerging and Selected Topics in Power Electronics, 3*(1), 147–162.

49. Mohamed, N., Aymen, F., Alharbi, T. E., El-Bayeh, C. Z., Lassaad, S., Ghoneim, S. S., & Eicker, U. (2022). A comprehensive analysis of wireless charging systems for electric vehicles. *IEEE Access, 10*, 43865–43881.

50. Neubauer, J., Pesaran, A., Bae, C., Elder, R., & Cunningham, B. (2014). Updating United States Advanced Battery Consortium and Department of Energy battery technology targets for battery electric vehicles. *Journal of Power Sources, 271*, 614–621.

51. Panchal, C., Stegen, S., & Lu, J. (2018). Review of static and dynamic wireless electric vehicle charging system. *Engineering Science and Technology, an International Journal, 21*(5), 922–937.

52. Patil, D., McDonough, M. K., Miller, J. M., Fahimi, B., & Balsara, P. T. (2017). Wireless power transfer for vehicular applications: Overview and challenges. *IEEE Transactions on Transportation Electrification, 4*(1), 3–37.

53. Peter, V. D. B., Tom, T., Noshin, O., & Joeri, V. M. (2016). Developments and challenges for EV charging infrastructure standardization. *World Electric Vehicle Journal, 8*(2), 557–563.

54. Prabhu, S. (2019). Wireless technology leading innovator—A case study of Qualcomm. *International Journal of Case Studies in Business, IT, and Education (IJCSBE), 3*(2), 68–78.

55. Rahimi-Eichi, H., Ojha, U., Baronti, F., & Chow, M. Y. (2013). Battery management system: An overview of its application in the smart grid and electric vehicles. *IEEE Industrial Electronics Magazine, 7*(2), 4–16.

56. Ramezani, A., Farhangi, S., Iman-Eini, H., Farhangi, B., Rahimi, R., & Moradi, G. R. (2018). Optimized LCC-series compensated resonant network for stationary wireless EV chargers. *IEEE Transactions on Industrial Electronics, 66*(4), 2756–2765.
57. Sadeghian, O., Oshnoei, A., Mohammadi-Ivatloo, B., Vahidinasab, V., & Anvari-Moghaddam, A. (2022). A comprehensive review on electric vehicles smart charging: Solutions, strategies, technologies, and challenges. *Journal of Energy Storage, 54*, 105241.
58. Sanguesa, J. A., Torres-Sanz, V., Garrido, P., Martinez, F. J., & Marquez-Barja, J. M. (2021). A review on electric vehicles: Technologies and challenges. *Smart Cities, 4*(1), 372–404.
59. Sasikala, G., & Ranjan, R. (2023). Wireless power transfer for IoT applications—A review. *Self-Powered Cyber Physical Systems*, 115–139.
60. Schmutzler, J., Andersen, C. A., & Wietfeld, C. (2013). Evaluation of OCPP and IEC 61850 for smart charging electric vehicles. *World Electric Vehicle Journal, 6*(4), 863–874.
61. Shareef, H., Islam, M. M., & Mohamed, A. (2016). A review of the state-of-the-art charging technologies, placement methodologies, and impacts of electric vehicles. *Renewable and Sustainable Energy Reviews, 64*, 403–420.
62. Shen, M., & Gao, Q. (2019). A review on battery management system from the modeling efforts to its multiapplication and integration. *International Journal of Energy Research, 43*(10), 5042–5075.
63. Song, K., Yang, G., Guo, Y., Lan, Y., Dong, S., Jiang, J., & Zhu, C. (2020). Design of DD coil with high misalignment tolerance and low EMF emissions for wireless electric vehicle charging systems. *IEEE Transactions on Power Electronics, 35*(9), 9034–9045.
64. Sulejmani, E., Beltle, M., & Tenbohlen, S. (2023). EMC of inductive automotive charging systems according to standard SAE J2954. *Vehicles, 5*(4), 1532–1552.
65. Sun, L., Ma, D., & Tang, H. (2018). A review of recent trends in wireless power transfer technology and its applications in electric vehicle wireless charging. *Renewable and Sustainable Energy Reviews, 91*, 490–503.
66. Sun, P., Bisschop, R., Niu, H., & Huang, X. (2020). A review of battery fires in electric vehicles. *Fire Technology, 56*(4), 1361–1410.
67. Szumanowski, A., & Chang, Y. (2008). Battery management system based on battery nonlinear dynamics modeling. *IEEE Transactions on Vehicular Technology, 57*(3), 1425–1432.

68. Uddin, K., Dubarry, M., & Glick, M. B. (2018). The viability of vehicle-to-grid operations from a battery technology and policy perspective. *Energy Policy, 113*, 342–347.

69. Xie, H., Huang, R., Sun, H., Han, Z., Jiang, M., Zhang, D., & Wu, T. (2023). Wireless energy: Paving the way for smart cities and a greener future. *Energy and Buildings, 297*, 113469.

70. Xing, Y., Ma, E. W., Tsui, K. L., & Pecht, M. (2011). Battery management systems in electric and hybrid vehicles. *Energies, 4*(11), 1840–1857.

71. Xiong, R., Li, L., & Tian, J. (2018). Towards a smarter battery management system: A critical review on battery state of health monitoring methods. *Journal of Power Sources, 405*, 18–29.

72. Yang, G., Song, K., Huang, X., Wang, C., Huang, X., Li, J., & Zhu, C. (2021). Improved interoperability evaluation method for wireless charging systems based on interface impedance. *IEEE Transactions on Power Electronics, 36*(8), 8588–8592.

73. Yang, L., Li, X., Liu, S., Xu, Z., Cai, C., & Guo, P. (2019). Analysis and design of three-coil structure WPT system with constant output current and voltage for battery charging applications. *IEEE Access, 7*, 87334–87344.

74. Zhang, X., Kan, T., You, C., & Mi, C. (2016). Modeling and analysis of AC output power factor for wireless chargers in electric vehicles. *IEEE Transactions on Power Electronics, 32*(2), 1481–1492.

75. Zhang, Y., You, P., & Cai, L. (2018). Optimal charging scheduling by pricing for EV charging station with dual charging modes. *IEEE Transactions on Intelligent Transportation Systems, 20*(9), 3386–3396.

76. Zhao, J., Cai, T., Duan, S., Feng, H., Chen, C., & Zhang, X. (2016). A general design method of primary compensation network for dynamic WPT system maintaining stable transmission power. *IEEE Transactions on Power Electronics, 31*(12), 8343–8358.

CHAPTER 4

STATIC AND DYNAMIC WIRELESS POWER TRANSFER FOR ELECTRIC VEHICLES

LEARNING OBJECTIVES

At the end of this chapter, readers will be able to:

1. Understand the components of a WPT system for EVs.
2. Understand principles and applications of magnetic resonance in WPT.
3. Recognize the working principles of inductive power transfer.
4. Compare and contrast different types of dynamic charging systems for EVs.
5. Explore the role of compensation circuits in wireless power transfer.

INTRODUCTORY EXAMPLE

Evatran, a leading company in electric vehicle technology, has developed innovative solutions for wireless charging. Picture yourself driving through a city in your electric vehicle, low on battery. Your dashboard alerts you of the situation. Luckily, Evatran's dynamic wireless charging station is nearby. Without the hassle of cables, your vehicle connects seamlessly to the station, replenishing its energy as you continue your journey.

This scenario is not fiction; it's the reality shaped by companies like Evatran, pioneers in wireless power transfer. Evatran's Plugless static charging system has revolutionized home and parking lot charging for electric vehicles, offering a convenient alternative to traditional methods. Moreover, Evatran's dynamic charging systems, developed with automotive manufacturers, promise even greater advancements.

By studying the concepts and uses of static and dynamic wireless power transfer, as represented by firms like Evatran, one can observe the emergence of a future in transportation that is powered wirelessly, effortlessly, and efficiently.

UNIT INTRODUCTION

The dynamic approach for wireless EV battery charging exclusively utilizes the inductive method for power transfer, while the static methodology for wireless EV battery charging uses both inductive and capacitive methods. There are two varieties of power transfer systems for electric vehicles, which are magnetic field coupling-based and electric field-based, depending on the transmission mechanism. The terms "inductive power transfer (IPT) system" and "capacitive power transfer (CPT) system" are alternative names for these systems (Yang et al., 2020).

This chapter examines static and dynamic wireless power transfer for electric vehicles, discussing concepts such as the physical principles of wireless power transfer, magnetic resonance, electromagnetic induction, microwaves, and electric coupling. The chapter also explores the static wireless power transfer system, providing a comprehensive explanation of the working principles of inductive power transfer, compensation circuits and coils, and configuration for capacitive power transfer. Furthermore, it examines the dynamic wireless power transfer system, which includes a few electric vehicles dynamic charging system types and carries out three dynamic charging systems comparisons (Faveto et al., 2022).

4.1. COMPONENTS OF WPT SYSTEM

Learning Objectives:

- Understand the components and functionality of Wireless Power Transfer (WPT) systems.

As seen in Figure 4.1, a wireless power transfer (WPT) system for electric vehicles (EVs) usually consists of two separate parts: the primary side and secondary side. The primary side has several components, such as a primary power converter, DC/AC power source, a primary controller, primary compensators, and a transmitting device, including metal plate (for capacitive power transfer) or a coil (for inductive power transmission). A secondary power converter, a receiving device (metal plate or coil), a secondary compensator, a secondary controller, and a DC load (battery) make up the secondary side (He & Guo, 2020).

Furthermore, a communication link has been integrated to enable the primary and secondary sides to exchange control commands and system status. A protection module that incorporates living objection detection (LOD) and foreign object detection (FOD) is a crucial part of the WPT system's safety features.

The utility grid receives its charging power from the primary power source. For the AC/DC power supply to comply with power-factor correction, grid interface regulations, or PFC, is required. Its objective is to minimize harmonic distortion on the input side and obtain a power factor of unity. The primary components are a high-frequency (HF) inverter, which transforms DC power into AC power, and an optional DC/DC converter, which controls the charging power (Zhao et al., 2016).

Certain configurations of wireless power transfer (WPT) systems control the voltage or charging current on the receiving end, eliminating the need for a DC/DC converter on the transmitting end. The high-frequency (HF) inverter on the primary side has a significant impact on the dimensions and effectiveness of the receiving as well as transmitting device. The HF inverter utilizes a soft-switching approach, ensuring optimal power use and stable performance. The method successfully lowers power losses because during the on/off transitions, the voltage or current of each switch is reduced to almost zero (Mai et al., 2017).

An HF rectifier and a DC/DC converter are optional add-ons for the secondary power converter. To provide power to a DC demand, such as the traction battery in an electric vehicle (EV), the HF rectifier transforms the AC voltage into DC voltage. There are two types of HF rectifiers: passive and active-controlled. The decision between the two options depends on the necessity of bidirectional power flow. By manipulating the direct current (DC) voltage at the output of the high frequency (HF) rectifier, it is possible to control the charging power in certain circumstances, hence reducing the requirement for the primary DC/DC converter (Li et al., 2021).

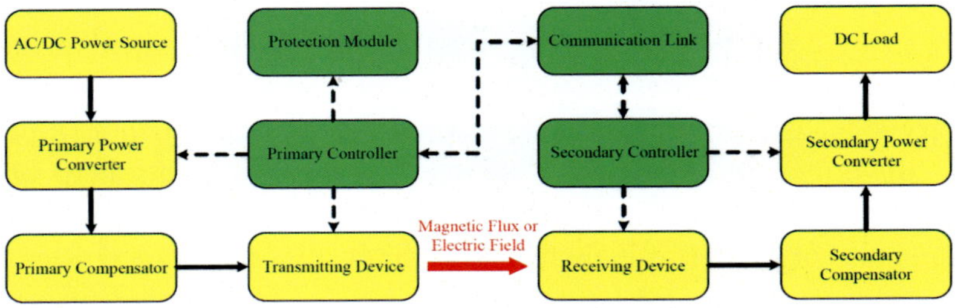

Figure 4.1. *Illustration of WPT system components for EVs.*

Source: Shubhangi Das; Creative Commons License.

Resonant LC circuits are commonly used as primary and secondary compensators to enhance power efficiency. In electric vehicle (EV) applications, the transmitting and receiving devices normally lie 15–30 cm apart, hence the coupling effect between them is usually minimal.

The secondary and primary compensators are essential in reducing the substantial reactive power of the HF inverter, which is necessary to stimulate the electric or magnetic field between the receiving and transmitting sides (Wang et al., 2017).

Therefore, increasing the primary high frequency (HF) inverter's power factor can lead to increased transmission efficiency. Additionally, the WPT system enables the generation of output characteristics that can be either constant voltage (CV) or constant current (CC) by utilizing different combinations of resonant inductors and capacitors. This function significantly simplifies the management of battery chargers for electric cars (EVs).

In WPT systems, the input and output power are regulated by the primary and secondary controllers. Monitoring the voltages and currents of the DC-link at both the primary and secondary sides require transmitting information wirelessly to the controllers. For real-time control in WPT systems, dependable and low-latency communication lines are essential due to the separation of power components. This ensures that the system can effectively handle various scenarios such as short circuit, no load, under voltage, open circuit, in-rush current, over voltage, FOD/LOD signals and transient over-shoot. It is preferable to manage protection and charging power control on one side, even though disconnecting the communication connections could cause problems and inefficiencies.

High levels of power can be transferred effectively between transmitting and receiving devices even in the presence of lateral and vertical misalignment. In order to enhance the efficiency of coils used in inductive power transfer (IPT) and metal plates used in capacitive power transfer (CPT), it is crucial to increase the mutual interactions between the electric field and magnetic flux, respectively. This can be accomplished by taking into account the dimensions of the primary and secondary devices.

One important factor to consider in the design of magnetic or electric couplers is the ability to maintain consistent mutual couplings even when experiencing lateral misalignment. This type of misalignment commonly occurs in EV applications, particularly in parking spaces. Later chapters will go into further detail on the topological design and construction of the magnetic couplers. This will help reduce the coupling variations that may arise due to lateral misalignment between the secondary and primary coils (Liu et al., 2021).

Figure 4.2 shows the arrangement of a standard IPT system designed for electric vehicle (EV) applications. As per the Standard SAE J2954 for light-duty EV/HEVs, the primary and secondary converters typically function within the frequency range of 80.0–90.0 kHz. A soft start feature and charging power management are made possible by the addition of a Buck converter to the main power converter stage (Mohamed et al., 2017).

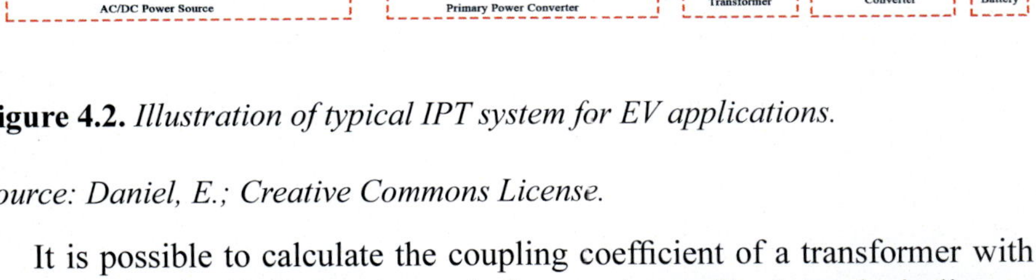

Figure 4.2. *Illustration of typical IPT system for EV applications.*

Source: Daniel, E.; Creative Commons License.

It is possible to calculate the coupling coefficient of a transformer with a loose coupling by observing the vertical separation and horizontal misalignment of the coils. The transmitting forms this type of transformer and receiving coils. To decrease electromagnetic interference (EMI) radiation and improve magnetic coupling, ferrite cores are frequently utilized in the construction of transmitting and receiving coils. However, this usage leads to iron losses and an increase in the overall mass of the system (Lu et al., 2016).

In order to recharge the batteries of electric and hybrid vehicles, the HF diode rectifier is frequently utilized as the primary power source for converting alternating current (AC) voltage into direct current (DC) voltage. In certain cases, the dual active bridge (DAB) design can be used as a replacement for the diode rectifier to provide bidirectional wireless power transmission between the grid and vehicle, as shown in vehicle-to-grid (V2G) applications (Peng et al., 2019).

4.2. WIRELESS POWER TRANSFER PHYSICAL PRINCIPLES

Learning Objectives:

- Understand the classification of Wireless Power Transfer (WPT) systems based on transmission principles.
- Explain the working principles and applications of electromagnetic induction in WPT.

Noncontact electrical transmission techniques are covered in this section. WPT systems can be categorized as magnetic resonance, electromagnetic induction, electrical resonance, microwaves, or electric coupling, based on the principles of transmission. Wireless Power Transfer systems can be categorized as microwaves, inductive coupling, lasers, magneto-dynamic coupling, capacitive coupling, and other combinations based on the differences between the transmitter and receiver. Wireless Power Transfer systems can be characterized as near-field or far-field techniques based on the transmission distance (Zhang et al., 2015).

In the near field or non-radiative zone, there is a regular exchange of energy between the source of radiation and the surrounding space. The far field, often referred to as the radiative region, is the area where energy is separated from the radiator and emitted as electromagnetic waves which move outward. There are different techniques used in wireless charging. Some of these techniques involve capacitive coupling and inductive coupling, which are known as near-field non-radiative techniques. On the other hand, there are far-field radiative techniques such as microwaves and lasers (Detka & Górecki, 2022).

4.2.1. Electromagnetic Induction

Transformers, which use a magnetic field to transfer power between mutually separated coils, are the most widely used electromagnetic induction-based devices. Figures 4.3 and 4.4 show the structures of the loosely coupled transformer and WPT system based on the electromagnetic induction.

The alternating current passing through L1 creates a varying magnetic field inside the magnetic core or silicon steel sheet, according to Ampere's theorem. The moving magnetic field then flows through the coil that receives it, and this is how Faraday's electromagnetic induction principle creates an alternating current (AC). The AC produced has the ability to power an AC load or can be transformed into a DC current through the rectifier to power a DC load (Doolittle & Brevik, 2014). The MAXWELL equations can be used to describe the concept of electromagnetic induction,

$$\nabla \times H = J + \frac{\partial D}{\partial t} \quad (4.1)$$

$$\nabla \times E = -\frac{\partial B}{\partial t} \quad (4.2)$$

Eqn. (4.1) demonstrates the relationship between the electric field and the magnetic field, while Eqn. (4.2) shows the reciprocal relationship between the electric and magnetic field (Egbert, 2002).

Electromagnetic induction remains a time-tested and extensively utilized technology for wireless power transfer. It has become increasingly popular in a wide range of commercial products, including razors and electric toothbrushes, effectively reducing the electric shock risk. Furthermore, electromagnetic induction based wireless power transfer (WPT) systems show great potential in various fields including electric vehicles, biomedicine, and portable devices like laptops and mobile phones (Thong & Gunstone, 2008).

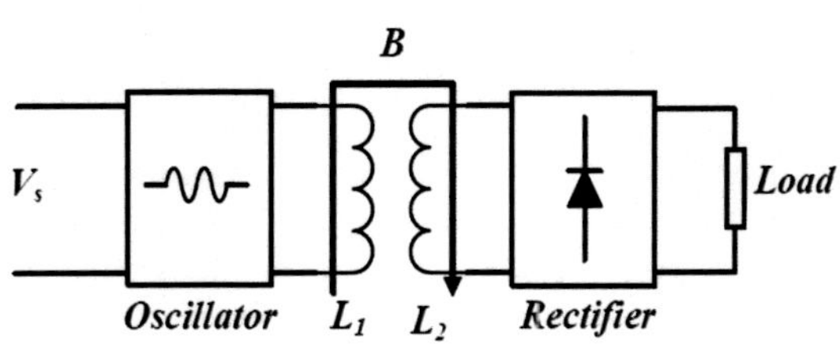

Figure 4.3. *Illustration of the WPT system's electromagnetic induction-based structure.*

Source: Naoki Shinora; Creative Commons License.

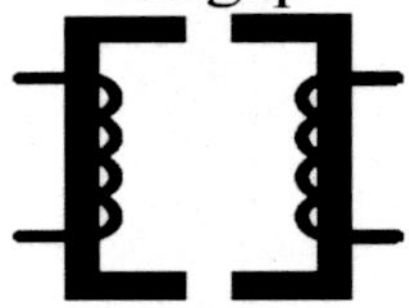

Figure 4.4. *Illustration of the loosely coupled transformer.*

Source: Benitto Albert; Creative Commons License.

Transmission power is proportional to coil frequency and mutual inductance. Mutual inductance refers to the magnetic flux generated by the transmitter coil as it passes through the receiving coil. When L_1 and L_2 are precisely positioned and placed very close to each other, the magnetic flux generated by L_1 passes through L_2 entirely. This results in the maximum value of mutual inductance and eliminates any loss in wireless transmission (Scanlon et al., 1969).

The coupling coefficient, typically denoted as 'k, $k = M/\sqrt{L_1 L_2}$ is frequently used to describe the level of coupling between the coils. To obtain maximum efficiency in power transmission, it is necessary to put the coils near each other. The mutual inductance is influenced by factors such as the coils' geometry and the presence of magnetic cores. Using bipolar structures and large, flat coil forms improves the coupling coefficient. The ferrite material boosts the coupling coefficient and reduces the divergence loss of magnetic field lines by effectively restricting the magnetic field (Chave & Booker, 1987).

A loosely connected transformer is also an integral part of an inductive power transfer (IPT) system. A large leakage inductance in a loosely coupled transformer can produce a lot of reactive power, which will reduce the efficiency and power transmission capacity of the system's circuit components. Hence, it is necessary to incorporate compensation circuits to decrease the apparent power demand of the power supply. Frequently used compensation circuits include single-stage (SS), split-phase (SP), push-pull (PP), series-parallel (PS), low-pass LC (LCL), and low-pass capacitor-capacitor (LCC) circuits, among others. Compensation circuits ensure that the input impedance of the system is completely resistive, allowing the system to function with a phase angle of zero (ZPA) (Hermance, 1983).

4.2.2. Magnetic Resonance

Conventional wireless power transfer (WPT) systems that depend on electromagnetic induction have a limited range for optimal efficiency and consistent power delivery. Magnetic resonance is used in current IPT systems to enhance both the efficiency and range of transmission. Tesla presented the concept of a magnetic resonance-based wireless power transfer (WPT) system a century ago. Figure 4.5 shows the experimental apparatus used for wireless power delivery (Dill, 2008).

A high frequency alternating current power supply, a spark gap switch (S.G.), a transmitting coil (P), a transformer, and a capacitor (C) make up the transmitter. Conversely, the receiver is made up of a 40-watt bulb and a receiving coil S. It has been discovered that the secondary side bulb may be illuminated by using an L-C resonant circuit on the main side. An L-C resonant circuit at the secondary side with a matched frequency can be added to increase the illumination range of the lightbulb. However, it should be noted that the increase in distance is still somewhat limited (Geva, 2006).

The concept of magnetic field resonance was initially introduced by MIT professor Marin Soljacic in 2007. It allows both the transmitter and the receiver to vibrate at the same frequency, facilitating effective energy transfer over greater distances. The research team conducted an experiment that demonstrated the ability of two copper coils, both with a radius of 25 cm, to successfully power a 60 W bulb. This was achieved at a distance of 2 m, which is eight times the radius of the coils. The frequency used for this experiment was 9.9 MHz, and the system efficiency was found to be approximately 40% (Katti & Shireen, 2011). The findings of his research greatly enhance the transmission distance of IPT systems (Figure 4.6).

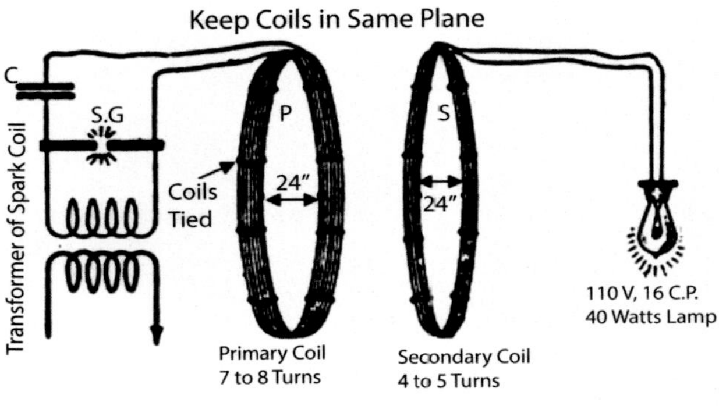

Figure 4.5. *Tesla's experimental wireless power supply device.*

Source: Suli Zou; Creative Commons License.

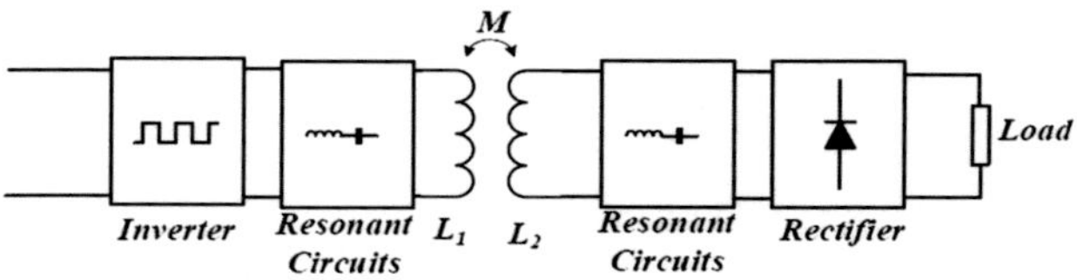

Figure 4.6. *The structure of WPT system based on magnetic resonance.*

Source: Zhongjing Ma; Creative Commons License.

This phenomenon effectively utilizes the "strong-coupled" regime in coupled-mode theory. In the transient (non-lossy) region close to the source, the resonators' high-quality factor (Q) results in a significantly faster rate of energy transfer compared to the loss caused by internal damping. Therefore, when comparing magnetic induction to magnetic resonance, it is observed that magnetic resonance can transmit the same amount of power across longer distances and with weaker magnetic fields, all while maintaining high efficiency (Matthews & Jezzard, 2004). The term 'mid-range' refers to a range that is 4–10 times the diameter of the coil, whereas the term 'short-range' refers to a range that is almost equal to the coil diameter. Magnetic resonance can harness the sub-wavelength dimensions of resonant coils to efficiently transmit energy, whereas magnetic induction can only utilize coils with dimensions on the scale of the wavelength to transmit power. As a result, magnetic resonance is a transfer method that functions at a moderate distance, while magnetic induction is a transfer method that functions at a close distance (Robitaille et al., 1998).

Practice Problem

Compare and contrast the operating principles of magnetic resonance and electric coupling in wireless power transfer systems, highlighting their advantages and limitations.

Solutions to Practice Problem

Magnetic resonance relies on the resonant coupling between coils to transfer energy efficiently over longer distances, whereas electric coupling requires close proximity between coils for efficient power transfer. Magnetic resonance offers greater flexibility in coil placement but may suffer from higher losses compared to electric coupling.

4.2.3. Electric Coupling

It is a technique which utilizes electric fields to directly transmit energy. Figure 4.7 shows the presence of two receiver plates and two transmitter plates positioned between the rectifier and oscillator. The transmitter plate and reception plate are aligned with one other, with an air space between them. Consequently, two capacitors (C_1 and C_2) are created, with air serving as the dielectric material. When a high-frequency alternating voltage is applied to the transmitter plates of C_1 and C_2, an alternating potential of the same frequency is generated by the oscillating electric field by electrostatic induction (Katsarakis et al., 2004).

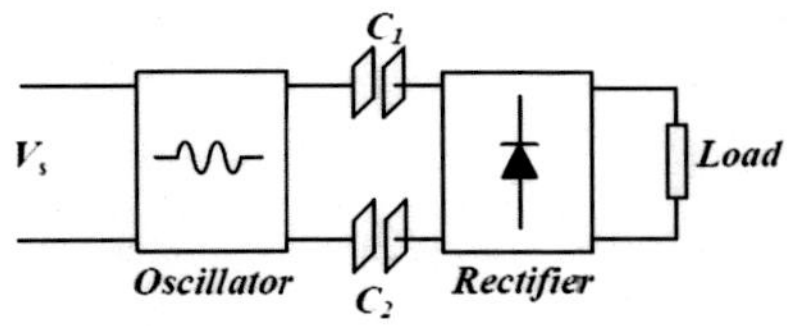

Figure 4.7. *The structure of the WPT system based on electric coupling.*

Source: Joa˜o, A.; Creative Commons License.

It is crucial to note that the voltage phases applied to C_1 and C_2 transmitters must be reversed. Furthermore, the phase of the generated alternating potentials in every transmitter plate must be directly opposed to one another in the receiving plate. The square of the voltage and capacitance has a direct relationship with the transferred power. The size of the plate and the separation between the transmitter and recipient plates determine the capacitance (Trif et al., 2008).

When evaluating various coupling methods, it is crucial to consider that electric coupling can transmit lower levels of power and may lead to high-voltage electric shock and the generation of harmful ozone. Electric coupling has found practical applications in various fields, including biomedical implants, portable devices, and integrated circuits (Harris, 2018).

4.3. THE STATIC WIRELESS POWER TRANSFER SYSTEM

Learning Objectives:

- Understand the distinction between static and dynamic wireless power transfer.
- Comprehend the working principles and components of inductive power transfer systems.

Depending on the condition of the electric vehicles being charged, two types of wireless charging systems are available: dynamic wireless power transfer (DWPT) and static wireless power transfer (SWPT).

The DWPT system enables charging of the vehicle while it is in motion, thereby enabling electric vehicles to utilize smaller batteries. SWPT is a versatile solution that can be used in a variety of settings, including shopping malls, parking lots, residential areas, and more (Subudhi, 2020).

All of them are based on the magnetic field coupling principles, and the circuit design is consistent across systems. However, the difference resides in the configuration of the magnetic circuit. Based on the transmission mechanism, there are two types of SWPT for EVs: magnetic field coupling-based and electric field-based power transfer systems. Whereas the latter is called the capacitive power transfer (CPT) system, the former is also called the inductive power transfer (IPT) system. It is crucial to note that IPT is more commonly used in SWPT for EVs compared to CPT (Hutchinson et al., 2019).

4.3.1. Inductive Power Transfer Working Principles

An IPT system consists of two main components: the primary side and the secondary side, as shown in Figure 4.8. A metal plate (CPT) or primary coil (IPT), converters, a DC or AC power source, the primary resonant circuit, are all part of the primary side, which is positioned on the ground. The secondary side of the system consists of the secondary converters, secondary resonant circuit, the direct current (DC) or alternating current (AC) load, and either a capacitive power transfer (CPT) metal plate or an inductive power transfer (IPT) secondary coil. This secondary side is located within electric vehicles (EVs). The two components are connected either through magnetic coupling or electrical coupling using coils or plates (Mayordomo et al., 2013).

The two primary components of an IPT system are a power factor correction (PFC) device and an inverter operating in the few kilohertz to several megahertz frequency range. This range makes sure that the system produces the most power and is as efficient as possible. The PFC function facilitates the transformation of low-voltage AC power obtained from the power grid into high-voltage DC power. Afterwards, the high-voltage DC power is subjected to a conversion process, where it is transformed into AC power with an accurate frequency utilizing the high-frequency inverter. The primary compensation circuit and the coil are then both filled with this AC power, enabling the inductive coupling between the two coils to transfer power to the secondary sides (Schormans et al., 2018). Furthermore, the power from the alternating current (AC) is converted to direct current (DC) power by the secondary high-frequency rectifier. Depending on the battery's state, a DC-DC converter is utilized to change the DC current or voltage to a level appropriate for charging. This is how an IPT system functions on a fundamental level.

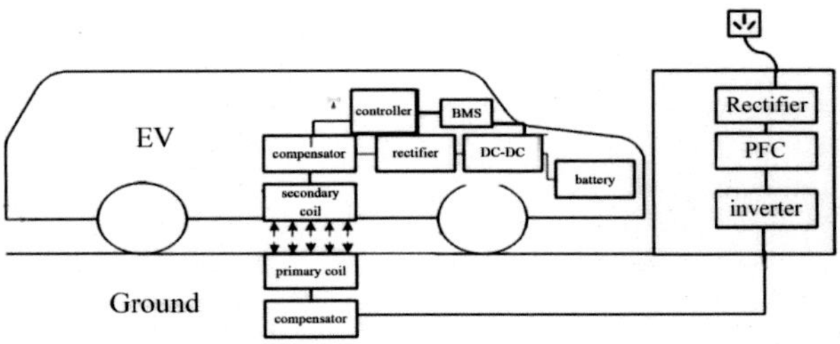

Figure 4.8. *Illustration of the overall configuration of WPT.*

Source: Pecas Lopes; Creative Commons License.

4.3.2. Compensation Coils and Circuits

Based on the working principle, it is evident that IPT transfers power through the use of inductive coupling between coils with a significant air gap. Hence, the efficacy of Inductive Power Transfer (IPT) diminishes with an increase in the gap between the primary and secondary coils. To address this issue, a compensation circuit is implemented to counteract the loss of magnetic flux caused by the increased gap (Wang et al., 2017).

The primary and secondary compensating circuits of the WPT system are mainly lossless L–C circuits. The coils' weakly connecting with the ground and the vehicle causes reactive power, which is reduced by these roles. Additionally, they also reduce the harmonic current and voltage. Integrating compensatory circuits into the charging system can substantially increase the power factor and drastically enhance transmission efficiency. Currently, there are different types of compensation circuits available, ranging from basic circuits to more complex ones such as LCC and SPS (Rovati & Cattini, 2011). Compensation circuits for wireless chargers can be designed using only two capacitors, which are connected either in parallel or series with the secondary and primary coils. Figure 4.9 shows the four topologies of compensating circuits: series-series (SS), series-parallel (SP), parallel-series (PS), and parallel-parallel (PP). Through the implementation of various compensation methods, the capacitance parameters of different topologies can be adjusted to function either as a voltage source or current source. Out of all the options available, the S-S topologies are highly favored for electric vehicle (EV) applications. This is mainly because of the variable coupling coefficient that proves to be beneficial in real-world scenarios (Zur & Stokar, 1996).

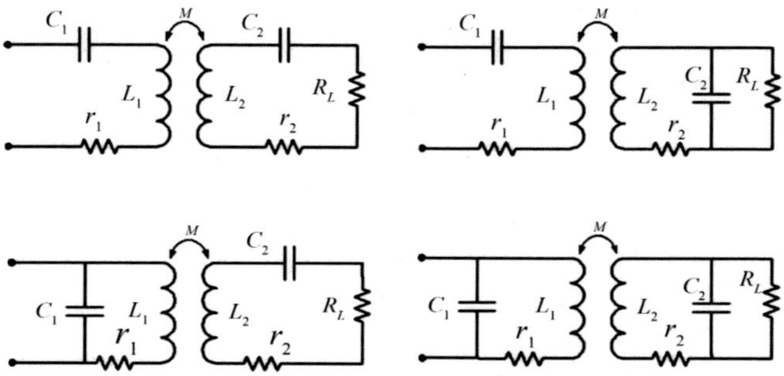

Figure 4.9. *Illustration of four compensation circuits with two capacitors.*

Source: Pecas Lopes; Creative Commons License.

Besides the compensation topologies mentioned earlier, there have been various proposed derived compensation circuits. Here is an example of the SPS compensation circuit, as shown in Figure 4.10. Another capacitor is utilized in the SPS compensation topology, allowing for a comparison with conventional topologies. This feature has the potential to enhance performance in scenarios where there is a lack of alignment between the primary and secondary coils (Sumner, 1947). There is another type of compensation topology known as LCC, which involves the use of an inductor and two capacitors, as shown in Figure 4.11.

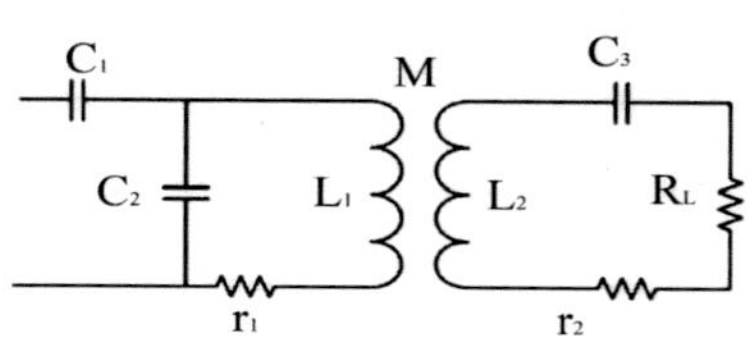

Figure 4.10. *Illustration of SPS compensation topology.*

Source: Francesco Marra; Creative Commons License.

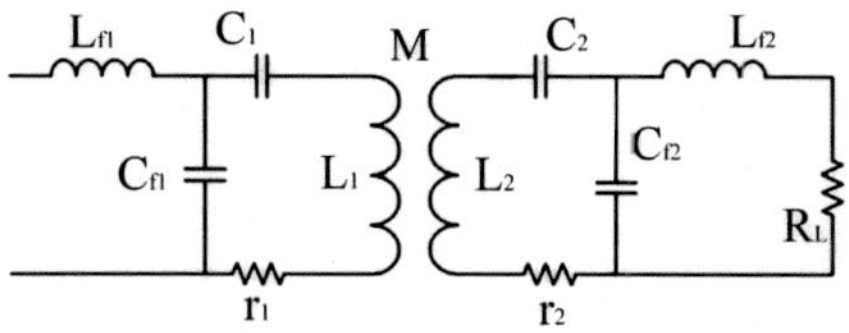

Figure 4.11. *Illustration of LCC compensation topology.*

Source: Garcia-Valle; Creative Commons License.

These topologies also differ in their compensatory strategies, which results in a wide range of properties. Transmission and reception coils are the two varieties of SWPT coils that are frequently used in the magnetic circuit of static wireless charging devices: square and circular. Double-D-quadrature (DDQ) coil and Double-D (DD) designs are the most common ones at the moment. Because the electric vehicle is immobile while being charged, the coupling coefficient remains constant.

4.3.3. Capacitive Power Transfer Configuration

One notable difference between the structure of a CPT system and that of an IPT system is the presence of secondary and primary plates, which are responsible for the generation of capacitors for power transmission. Unlike IPT, the CPT system uses electric fields rather than magnetic fields to facilitate power transfer. Nevertheless, CPT systems primarily prioritize applications within short distances of around 1 mm. This transfer distance falls significantly short of the ground clearance of electric vehicles (EVs), thus restricting its suitability for SWPT (wireless power transfer) in EVs (Erel et al., 2021).

4.4. THE DYNAMIC WIRELESS POWER TRANSFER SYSTEM

Learning Objectives:

- Understand the structure and operation of dynamic wireless charging systems.
- Compare different types of dynamic electric vehicle charging systems.

A CPT system is set up similarly to an IPT system, except for the primary and secondary plates that generate capacitors for power transfer. In contrast to IPT, CPT technology facilitates power transfer by using electric fields rather than magnetic fields. Nevertheless, CPT systems primarily prioritize applications within short distances of around 1 mm. This transfer distance falls significantly short of the ground clearance of electric vehicles (EVs), thus restricting its suitability for SWPT (wireless power transfer) in EVs (Li et al., 2016). Both the static wireless charging system and the dynamic wireless power transfer system use receiving coils that are comparable to one another. Nevertheless, there is a distinction in the configuration of the transmitting coil in both systems. Transmitting ends are classified by coil shape into three types: segmented rail type, array type, and long rail type (Tian et al., 2020).

4.4.1. Electric Vehicle Dynamic Charging System Types

An innovative approach to wireless charging in electric vehicles involves the use of coil arrays strategically placed along the vehicle's driving route. When an electric vehicle is in motion, the energy is seamlessly transferred to the on-board battery through electromagnetic coupling using a coil array beneath the road surface (Panchal et al., 2018). This technology eliminates the need for extensive or even any energy storage battery packs in electric vehicles. This approach enhances the driving distance of electric vehicles and enhances the ease and security of power provision. Various power transmission rails are utilized in dynamic wireless charging systems for electric vehicles. These include segmented rails and centralize long rails. Centralized long rails have a large power loss and poor transmission efficiency. Thus, in order to enhance power transmission efficiency, a more viable approach would be the implementation of multi-segment rail switching in wireless power supply technology (Moghaddam et al., 2019).

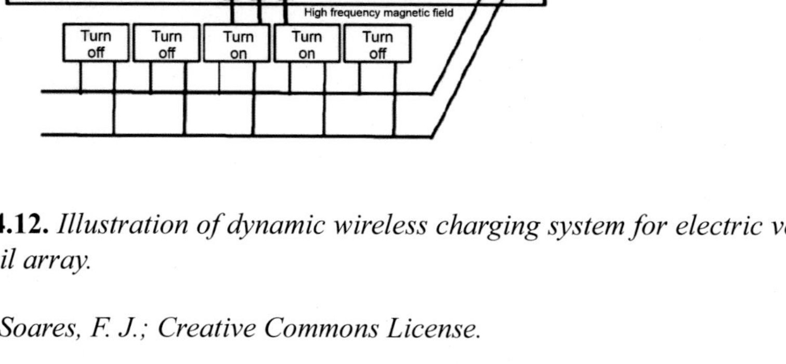

Figure 4.12. *Illustration of dynamic wireless charging system for electric vehicles using coil array.*

Source: Soares, F. J.; Creative Commons License.

The only rail that is activated underneath the car is the matching one. Once the vehicle departs, the adjacent section of the rail is automatically activated, while the power supply to the rail is promptly disconnected. Use the segmented guide rail's wireless power supply technology to locate the vehicle to activate or deactivate the power transmission guide rail. The current approach to locating electric vehicles involves the installation of a sensor on the guide rail or the attachment of additional signal transmission coils to both the car and the rail. These measures are taken to accurately identify and gather essential data (Hemavathi & Shinisha, 2022) (Figures 4.12–4.14).

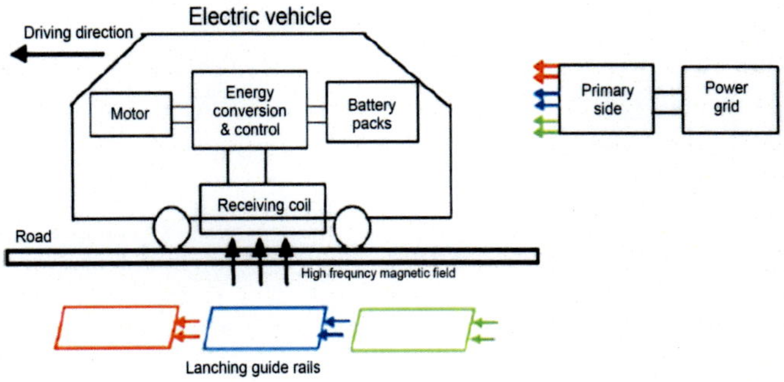

Figure 4.13. *Illustration of dynamic wireless charging system for electric vehicles using segmented rail.*

Source: Philip, B.; Creative Commons License.

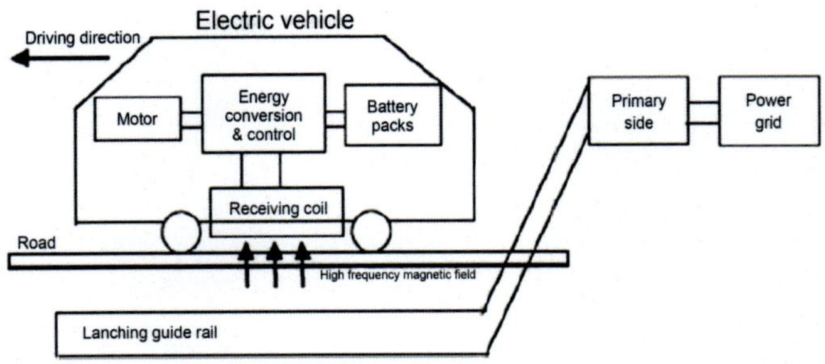

Figure 4.14. *Illustration of dynamic wireless charging system for electric vehicles using long rail.*

Source: Rocha, P. M.; Creative Commons License.

4.4.2. Three Dynamic Charging Systems Comparison

Table 4.1 shows the features of the three dynamic systems that were discussed. There are three types of dynamic wireless charging systems for electric vehicles, each with its own unique characteristics. One of these types is the long rail type, which is known for its single-stage rail design. If there is a problem with the rail, the entire line may become non-functional. However, one benefit is that it leads to reduced voltage fluctuations in the output (Lagnelöv et al., 2020).

The coil array type rail ensures a high level of dependability by allowing each coil to function independently. This means that the malfunction of a single guide rail does not have any effect on the coils of other guide rails. However, there are fluctuations in the output voltage, due to the various positions at which the coils switch. The segmented guide rail functions as a structure that effectively integrates the advantages of both components; while reducing any potential drawbacks they might pose to the overall system (Shanmugam et al., 2022).

Table 4.1. *Features of Three Types of Dynamic Charging Systems for EVs*

Classification	**Voltage Fluctuation**	**Coil Numbers and Charging Efficiency**	**Construction Cycle and Cost**
Segmented rail	There are multiple switching sites available, and the voltage at each switching position can fluctuate.	Large primary launching guide rail number, complex control, high system efficiency.	High cost and moderate difficulty in construction.
Coil array	At the switching position, there are numerous points of transition and a voltage that constantly varies.	Primary-side large number of transmit coils, complicated control, and high system efficiency.	High cost and long construction period.
Long rail	When using an electric vehicle, there is very little variation in the output voltage.	Single launching guide rail, easy control, and low system efficiency.	Low cost and short construction period.

Source: Chong Zhu, Creative Commons License.

SUMMARY

- The chapter examines the fundamental ideas and mechanisms of dynamic and static wireless power transmission for electric cars. It begins by providing a comprehensive introduction to the fundamental principles of magnetic resonance, electromagnetic induction, microwaves, and electric coupling.
- The next section covers static wireless power transfer systems, with a focus on inductive power transmission. It includes a description of the coil layouts, compensation circuits, and operating principles. Configurations for capacitive power transfer are also described.
- The chapter focuses on dynamic wireless power transfer and examines different kinds of electric vehicle dynamic charging systems. It also includes a detailed comparative analysis of three specific dynamic charging systems.

REVIEW QUESTIONS

1. Explain wireless power transfer principles in EV charging.
2. Compare static and dynamic WPT systems for EVs.
3. Describe components in static WPT for EVs.
4. Discuss compensation circuits in inductive WPT.
5. Evaluate dynamic WPT vs. traditional charging.
6. Identify types of dynamic EV charging systems.

MULTIPLE CHOICE QUESTIONS

1. Which of the following is NOT a principle of wireless power transfer?

a. Electromagnetic induction

b. Magnetic resonance

c. Direct current

d. Electric coupling

2. What is the primary focus of static wireless power transfer systems?

a. Magnetic resonance

b. Capacitive power transfer

c. Inductive power transfer

d. Microwave transmission

3. **Which component is essential in compensating for power loss in inductive power transfer?**
   a. Capacitor
   b. Resistor
   c. Compensation circuit
   d. Transformer
4. **Dynamic wireless power transfer systems are primarily designed for:**
   a. Stationary charging
   b. Charging while in motion
   c. Long-distance charging
   d. Remote charging
5. **Which of the following is NOT a type of electric vehicle dynamic charging system?**
   a. Capacitive charging system
   b. Inductive charging system
   c. Overhead charging system
   d. Wireless charging system

Answers to Multiple Questions

1. (c); 2. (c); 3. (c); 4. (b); 5. (a)

REFERENCES

1. Chave, A. D., & Booker, J. R. (1987). Electromagnetic induction studies. *Reviews of Geophysics, 25*(5), 989–1003.
2. Detka, K., & Górecki, K. (2022). Wireless power transfer—A review. *Energies, 15*(19), 7236.
3. Dill, T. (2008). Contraindications to magnetic resonance imaging. *Heart, 94*(7), 943–948.
4. Doolittle, J. A., & Brevik, E. C. (2014). The use of electromagnetic induction techniques in soils studies. *Geoderma, 223*, 33–45.
5. Egbert, G. D. (2002). Processing and interpretation of electromagnetic induction array data. *Surveys in Geophysics, 23*(2), 207–249.
6. Erel, M. Z., Bayindir, K. C., Aydemir, M. T., Chaudhary, S. K., & Guerrero, J. M. (2021). A comprehensive review on wireless capacitive power transfer technology: Fundamentals and applications. *IEEE Access, 10*, 3116–3143.

7. Faveto, A., Panza, L., Bruno, G., Cirimele, V., Furio, S. S., & Lombardi, F. (2022). Efficient management of industrial electric vehicles by means of static and dynamic wireless power transfer systems. *The International Journal of Advanced Manufacturing Technology, 123*(3), 1249–1267.
8. Geva, T. (2006). Magnetic resonance imaging: Historical perspective. *Journal of Cardiovascular Magnetic Resonance, 8*(4), 573–580.
9. Harris, A. L. (2018). Electrical coupling and its channels. *Journal of General Physiology, 150*(12), 1606–1639.
10. He, L., & Guo, D. (2020). An active switched-capacitor half-wave receiver with high efficiency and reduced components in WPT system. *IEEE Transactions on Industrial Electronics, 68*(12), 12119–12129.
11. Hemavathi, S., & Shinisha, A. (2022). A study on trends and developments in electric vehicle charging technologies. *Journal of Energy Storage, 52*, 105013.
12. Hermance, J. F. (1983). Electromagnetic induction studies. *Reviews of Geophysics, 21*(3), 652–665.
13. Hutchinson, L., Waterson, B., Anvari, B., & Naberezhnykh, D. (2019). Potential of wireless power transfer for dynamic charging of electric vehicles. *IET Intelligent Transport Systems, 13*(1), 3–12.
14. Katsarakis, N., Koschny, T., Kafesaki, M., Economou, E. N., & Soukoulis, C. M. (2004). Electric coupling to the magnetic resonance of split ring resonators. *Applied Physics Letters, 84*(15), 2943–2945.
15. Katti, G., Ara, S. A., & Shireen, A. (2011). Magnetic resonance imaging (MRI)–A review. *International Journal of Dental Clinics, 3*(1), 65–70.
16. Lagnelöv, O., Larsson, G., Nilsson, D., Larsolle, A., & Hansson, P. A. (2020). Performance comparison of charging systems for autonomous electric field tractors using dynamic simulation. *Biosystems Engineering, 194*, 121–137.
17. Li, S., Liu, Z., Zhao, H., Zhu, L., Shuai, C., & Chen, Z. (2016). Wireless power transfer by electric field resonance and its application in dynamic charging. *IEEE Transactions on Industrial Electronics, 63*(10), 6602–6612.
18. Li, W., Zhang, Q., Cui, C., & Wei, G. (2021). A self-tuning S/S compensation WPT system without parameter recognition. *IEEE Transactions on Industrial Electronics, 69*(7), 6741–6750.
19. Liu, S., Zhao, X., Wu, Y., Zhou, L., Li, Y., Mai, R., & He, Z. (2021). Efficiency improvement of dual-receiver WPT systems based on partial power processing control. *IEEE Transactions on Power Electronics, 37*(6), 7456–7469.

20. Lu, F., Zhang, H., Hofmann, H., & Mi, C. C. (2016). An inductive and capacitive combined wireless power transfer system with LC-compensated topology. *IEEE Transactions on Power Electronics, 31*(12), 8471–8482.
21. Mai, R., Chen, Y., Zhang, Y., Yang, N., Cao, G., & He, Z. (2017). Optimization of the passive components for an S-LCC topology-based WPT system for charging massive electric bicycles. *IEEE Transactions on Industrial Electronics, 65*(7), 5497–5508.
22. Matthews, P. M., & Jezzard, P. (2004). Functional magnetic resonance imaging. *Journal of Neurology, Neurosurgery & Psychiatry, 75*(1), 6–12.
23. Mayordomo, I., Dräger, T., Spies, P., Bernhard, J., & Pflaum, A. (2013). An overview of technical challenges and advances of inductive wireless power transmission. *Proceedings of the IEEE, 101*(6), 1302–1311.
24. Moghaddam, Z., Ahmad, I., Habibi, D., & Masoum, M. A. (2019). A coordinated dynamic pricing model for electric vehicle charging stations. *IEEE Transactions on Transportation Electrification, 5*(1), 226–238.
25. Mohamed, A. A., Lashway, C. R., & Mohammed, O. (2017). Modeling and feasibility analysis of quasi-dynamic WPT system for EV applications. *IEEE Transactions on Transportation Electrification, 3*(2), 343–353.
26. Panchal, C., Stegen, S., & Lu, J. (2018). Review of static and dynamic wireless electric vehicle charging system. *Engineering Science and Technology, an International Journal, 21*(5), 922–937.
27. Peng, C., Chen, Z., Liu, Z., Wang, J. F., Liang, J., Li, C., & Li, J. (2019). On the load-independence of a multi-receiver wireless power transfer system. *IEEE Microwave and Wireless Components Letters, 29*(8), 563–565.
28. Robitaille, P. M., Abduljalil, A. M., Kangarlu, A., Zhang, X., Yu, Y., Burgess, R., & Spigos, D. (1998). Human magnetic resonance imaging at 8 T. *NMR in Biomedicine: An International Journal Devoted to the Development and Application of Magnetic Resonance In Vivo, 11*(6), 263–265.
29. Rovati, L., & Cattini, S. (2011). Zero-field readout electronics for planar fluxgate sensors without compensation coil. *IEEE Transactions on Industrial Electronics, 59*(1), 571–578.
30. Scanlon, P. J., Henriksen, R. N., & Allen, J. R. (1969). Approaches to electromagnetic induction. *American Journal of Physics, 37*(7), 698–708.
31. Schormans, M., Valente, V., & Demosthenous, A. (2018). Practical inductive link design for biomedical wireless power transfer: A tutorial. *IEEE Transactions on Biomedical Circuits and Systems, 12*(5), 1112–1130.

32. Shanmugam, Y., Narayanamoorthi, R., Vishnuram, P., Bajaj, M., AboRas, K. M., & Thakur, P. (2022). A systematic review of dynamic wireless charging system for electric transportation. *IEEE Access, 10*, 133617–133642.
33. Subudhi, P. S. (2020). Wireless power transfer topologies used for static and dynamic charging of EV battery: A review. *International Journal of Emerging Electric Power Systems, 21*(1), 20190151.
34. Sumner, J. H. (1947). The theory and operation of Petersen coils. *Journal of the Institution of Electrical Engineers-Part II: Power Engineering, 94*(40), 283–298.
35. Thong, W. M., & Gunstone, R. (2008). Some student conceptions of electromagnetic induction. *Research in Science Education, 38*, 31–44.
36. Tian, Y., Zhu, Z., Xiang, L., & Tian, J. (2020). Vision-based rapid power control for a dynamic wireless power transfer system of electric vehicles. *IEEE Access, 8*, 78764–78778.
37. Trif, M., Troiani, F., Stepanenko, D., & Loss, D. (2008). Spin-electric coupling in molecular magnets. *Physical Review Letters, 101*(21), 217201.
38. Wang, S., Chen, J., Hu, Z., Rong, C., & Liu, M. (2017). Optimization design for series–Series dynamic WPT system maintaining stable transfer power. *IET Power Electronics, 10*(9), 987–995.
39. Wang, Y., Yao, Y., Liu, X., Xu, D., & Cai, L. (2017). An LC/S compensation topology and coil design technique for wireless power transfer. *IEEE Transactions on Power Electronics, 33*(3), 2007–2025.
40. Yang, L., Li, X., Liu, S., Xu, Z., & Cai, C. (2020). Analysis and design of an LCCC/S-compensated WPT system with constant output characteristics for battery charging applications. *IEEE Journal of Emerging and Selected Topics in Power Electronics, 9*(1), 1169–1180.
41. Zhang, C., Lin, D., & Hui, S. Y. (2015). Basic control principles of omnidirectional wireless power transfer. *IEEE Transactions on Power Electronics, 31*(7), 5215–5227.
42. Zhao, J., Cai, T., Duan, S., Feng, H., Chen, C., & Zhang, X. (2016). A general design method of primary compensation network for dynamic WPT system maintaining stable transmission power. *IEEE Transactions on Power Electronics, 31*(12), 8343–8358.
43. Zur, Y., & Stokar, S. (1996). An algorithm for eddy currents summarization and compensation. *Magnetic Resonance in Medicine, 35*(2), 252–260.

CHAPTER 5

TECHNOLOGICAL ADVANCES IN ELECTRIC VEHICLES

LEARNING OBJECTIVES

At the end of this chapter, readers will be able to:

1. Understand the historical evolution and significance of electric vehicles.
2. Explore the development status and market trends of electric vehicles globally.
3. Know the principles and advantages of wireless charging technologies for electric vehicles.
4. Explore the evolution of EV charging technologies, including different charging levels and strategies.
5. Understand the key challenges and solutions in wireless power transfer (WPT) for electric vehicles (EVs).
6. Identify the role of EVs in the energy Internet and their impact on new energy micro-grid systems.

INTRODUCTORY EXAMPLE

Tesla, Inc., established by Elon Musk in 2003, serves as a prime illustration of the progress within the electric vehicle (EV) industry. With the production of the Tesla Roadster, the company continues to push the limits of electric vehicle technology, achieving significant advancements in range, battery efficiency, and performance. The emergence of models including the Model S, Model X, Model 3, and Model Y has not just only transformed the automotive market but has also redefined the way electric vehicles are viewed as viable options to conventional gasoline-powered cars.

Tesla's dedication to pushing boundaries goes far beyond just the aesthetics and capabilities of its vehicles. The company's extensive Supercharger network has effectively alleviated one of the primary concerns for electric vehicle drivers the fear of running out of battery power. With thousands of charging stations strategically placed around the globe, EV drivers can now enjoy fast and convenient charging solutions. In addition, the progress made by Tesla in the field of autonomous driving technology, exemplified by the Autopilot feature, has set the stage for a future of transportation that is both safer and more efficient.

In the ever-evolving landscape of electric vehicles, Tesla continues to lead the way, pushing boundaries and paving the path toward a more sustainable future in transportation. Tesla's bold vision for growth is evident in its plans for constructing Giga factories and creating new vehicle models. Their commitment to a more sustainable electrified transport ecosystem sets them apart as a leader in the industry.

UNIT INTRODUCTION

Wireless power transmission for supplying electricity to electrical devices and equipment, as well as for charging, has been a topic of interest since the era of Nikola Tesla. Yet this was not feasible at that time due to the unavailability of the necessary enabling technologies. In 2007, researchers made a significant breakthrough by successfully lighting up a bulb using wireless power from a distance equal to two meters. Significant progress has been achieved in this area since this notable achievement. Wireless power transfer (WPT) is being actively explored in the field of electric vehicle (EV) charging due to its numerous advantages. This technology has great potential and is being studied extensively (Sanguesa et al., 2021).

Throughout this unit, the vast world of electrified transportation is explored, covering everything from the past development of electric vehicles through the cutting-edge advancements of today. One should start by exploring the origins and evolution of electric vehicles, which will help us grasp their present-day importance. Exploring the differences between Battery Electric Vehicles (BEVs) and Hybrid Electric Vehicles (HEVs), one analyzes their structures and capabilities. Interest should be taken in exploring the current state of electric vehicle development on a global scale, examining the progress made both domestically and internationally. Next, the exciting world of mobile charging technologies is explored, examining both static and dynamic methods and their potential impact on the future of electric vehicle infrastructure (Salkuti, 2023). In addition, the crucial role of batteries made of lithium-ion in fueling electric vehicles is examined and delve into the most recent developments in electric vehicle charging technologies. Shortly, one can expect to see exciting developments in the world of electric vehicles. Imagine parking lots equipped with multi-layer wireless charging technology and automated parking systems. These advancements are just a glimpse of the innovative ideas driving the electric vehicle revolution (Sharma et al., 2020).

5.1. OVERVIEW OF ELECTRIC VEHICLE DEVELOPMENT

Learning Objectives:

- Explore the concepts and advantages of wireless charging technologies for electric vehicles.
- Differentiate between static and dynamic wireless charging methods for electric vehicles.
- Evaluate the potential of wireless charging in revolutionizing EV infrastructure and user experience.

5.1.1. History of Electric Vehicles

Automobiles play a vital role in improving people's mobility, serving as an important form of transportation. With the improvement of people's living standards, there has been a significant increase in the number of cars on the road. However, with the rise of the oil crisis and growing concerns about polluting the environment, conventional fuel vehicles no longer align with the demands of sustainable development. On the other hand, electric vehicles have experienced significant growth thanks to their ability to conserve energy, reduce emissions, and protect the environment. In recent years, governments around the world have implemented many incentives to encourage the growth of electric vehicles. These measures have played a crucial role in advancing the development of this technology. Currently, the field of electric cars is experiencing significant growth on a global scale, with sales consistently increasing. Electric vehicles constitute the undeniable future of the automotive industry's progress (Rajashekara, 1994).

Back in 1832, the brilliant mind of Scottish inventor Robert Anderson revolutionized transportation by equipping a carriage with batteries and motors, creating the world's first electric carriage. In 1834, Thomas Davenport, an American inventor, revolutionized transportation with his introduction of an early electric tricycle. The invention, as described by Ajanovic (2015), ushered in a new era at transportation with its utilization of a dry battery as well a DC motor. It's worth noting that this development occurred much earlier than the invention of the internal combustion engine, which was first created in 1885 (Figure 5.1).

In April 1881, Gustave Truvi from France constructed a three-wheel electric automobile equipped with lead-acid batteries. And so, the initial version of electric vehicles came into existence. During that period, a wave of innovative electric vehicles emerged and gained popularity across Europe. These vehicles played a significant role in transportation during the latter part of the 19th and early 20th centuries. During the 1920s, the emergence of combustion engines propelled fuel vehicles to the forefront of transportation. On the other hand, electric vehicles tend to have limited speed and range, while traditional fuel vehicles boast impressive endurance and benefit from affordable gasoline prices. Traditional fuel vehicles have outperformed electric vehicles in various areas, leading to a gradual decline in the popularity of electric vehicles within the market (Climent et al., 2021).

Figure 5.1. *Illustration of davenport with his electric tricycle.*

Source: Triviño-Cabrera; Creative Commons License.

In the 1980s, there was a renewed interest in electric vehicles due to concerns about the oil crisis and environmental pollution. Several developed nations have started to reexamine electric vehicles. Electric vehicles have seen significant advancements thanks to the strong support and promotion from various countries, making them more accessible to the public (Kubański, 2020).

5.1.2. Electric Vehicles and Wireless Charging

Given the pressing global climate concerns and the growing shortage of petroleum, electric vehicles are effectively entering the automotive industry. Electric vehicles provide a more convenient method of charging and result in a significantly lower impact on the environment. With the continuous development of related technologies, electric vehicles have emerged as the leading players in the automobile market, capturing the hearts of people who are always on the move. Electric vehicle technology research encompasses various aspects, including power battery systems, vehicle control, charging systems, motor drive systems, and functional safety. One of the key factors that significantly impacts the electric vehicle's overall performance is the power battery, which serves as its core component. As power batteries become more prevalent, it is essential to focus on developing efficient recharging systems for electric vehicles to ensure they never run out of power (Ahmad et al., 2017).

Currently, there are two main types of charging techniques for electric vehicle power replacement or charging. The power replacement technique involves swapping out a depleted battery with a fully charged one. While this approach can be achieved, it encounters various challenges in terms of design, production, and marketing. Take, for instance, the challenges posed by the removable power battery in terms of safety, the need for standardized power battery requirements, and the worries of car owners. The charging method proves to be more viable when compared to power replacement. There are different methods available for charging power batteries, including both contact and non-contact charging options. Conductive charging is commonly referred to as contact charging. The charging process is facilitated through the physical connection between the wire plug and the charging pile, allowing for efficient transfer of electricity to the electric vehicle's charger socket. Wireless charging, a cutting-edge method of charging, eliminates the need for physical contact. This technique employs electromagnetic waves as a means for the transmission of electrical energy. When recharging an electric car, there is no requirement for a physical connection between the power supply gear on the ground with the vehicle's power reception equipment (Amjad et al., 2022). The technique of contact charging has reached a relatively advanced stage of development. Nevertheless, there are persisting issues such as the laborious process of connecting and disconnecting cables, the potential danger of an electrical shock, and the significant consequences resulting from harsh weather conditions like rain and snow. Wireless charging, as a nascent charging or plug-in technique, offers distinct advantages compared to the existing contact charging or technique in terms of charging station infrastructure, practical application, and subsequent upkeep: (i) enhanced adaptation to the environment and increased reliability through sealed installation; (ii) optimal space utilization, making it particularly well-suited for the construction of large-scale clusters in the future; (iii) convenient application and management, particularly well-suited for the future trend of unmanned management development; and (iv) ensuring a high level of safety by eliminating the danger of electrical shock (Figure 5.2) (Ou et al., 2014).

Figure 5.2. *Illustration of pure electric vehicle power system structure.*

Source: Chong Zhu; Creative Commons License.

There are two main types of wireless charging for electric vehicles: static charging and dynamic charging. Static charging refers to when the car is stationary during the charging process (Mohamed et al., 2022).

5.1.3. Static Wireless Charging

The exponential expansion of electric vehicles also indicates an extensive market in wireless charging technology. Electric transit buses and electric automobiles are commonly utilized in the realm of electric vehicles. Traditional methods of charging electric buses involve the use of a conductive DC power supply system. This system is designed to connect to a vehicle's charging socket utilizing a charging gun that is specifically made for non-vehicle chargers. Charging stations often necessitate a considerable amount of space.

Managing stations for charging can be expensive. There are significant issues with the vehicle and charging gun charging socket, resulting in problems like unreliable contact, overheating, and a limited lifespan for both components. Wireless charging technology offers a hassle-free and dependable method of charging, enhancing the overall charging experience for users and alleviating any inconvenience (Panchal et al., 2018).

In the concept of the futuristic city, autonomous cars have seamlessly integrated into the daily lives of people, serving as a convenient and reliable mode of transportation. In a few years, parking lots will be equipped with state-of-the-art wireless charging devices in each space. This innovative technology allows vehicles parked in the lot to be efficiently charged, ensuring that their cruising range remains optimal. Furthermore, a wireless charging station is poised to become a standard method of charging shortly. When it comes to electric vehicle charging, wireless charging offers significant advantages over traditional methods. It not only saves valuable floor space but also increases the rate of charging per unit area. This means that electric car owners can enjoy the convenience of faster and more efficient charging. In the future, the combination of wireless charging and automated parking technology holds great potential. Simply put, stationary wireless charging is a crucial building block for achieving the fully automated autonomous driving of electric vehicles (Lukic & Pantic, 2013).

5.1.4. Dynamic Wireless Charging

Wireless charging technology has emerged as a solution to the limitations and safety concerns associated with traditional conductive charging methods. Consequently, it has gained significant traction in electric vehicle recharging. However, both wired charging and static wireless charging face various challenges, including the requirement for frequent charging, limited driving range, substantial battery usage, and high costs. Electric buses have seen a surge in demand for longer battery life. The current static wireless charging infrastructure, including parking lots and charging stations, are facing challenges in meeting the growing demand for electric vehicles. Urban road charging is expected to have a substantial impact, with dynamic wireless charging emerging as the primary mode. Charging electric vehicles while they are in motion, known as dynamic charging, has become a significant development in the field. It has the potential to decrease the capacity of the car's battery and compromise the overall quality of the vehicle. However, it allows for the use of a smaller number of battery packs in electric vehicles, so increasing their range (Miller et al., 2014). In addition, the power supply gets safer and more convenient (Figure 5.3).

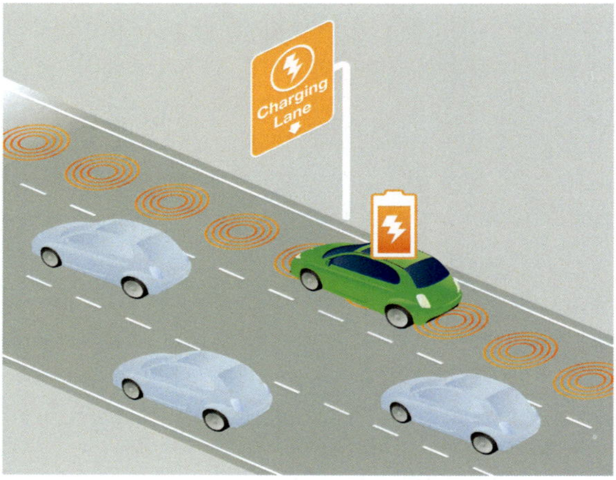

Figure 5.3. *Illustration of dynamic wireless charging scenario.*

Source: Chong Zhu; Creative Commons License.

Dynamic wireless chargers provide a crucial catalyst that ensures electric vehicles overcome the limitations of their cruising range and genuinely establish themselves as the dominant force in the market. Hence, the primary research emphasis of major research institutes both domestically and internationally is the creation of a dynamic wireless power supply system that possesses characteristics including increased power, enhanced safety measures, improved lateral adaptability, reduced electromagnetic radiation, heightened efficiency, and a reasonable cost (Zhang et al., 2016).

5.2. EV CHARGING TECHNOLOGIES AND LITHIUM-ION BATTERY

Learning Objectives:

- Understand the historical development and key innovations in lithium-ion battery technology.
- Analyze the evolution of EV charging technologies, including charging levels and strategies.
- Evaluate advanced charging algorithms and their impact on battery health.

5.2.1. The Lithium-Ion Battery Technology Development

In the 1970s, there was a surge in research on lithium-ion batteries, driven by the pressing energy crisis at the time. Anode materials in this study included Li or Li-Al alloys. Following that, a series of lithium primary batteries emerged, such as Li/MnO_2, Li/I_2, $Li/SOCl_2$, and Li/FeS_2. In 1970, Panasonic received a patent over Li/(CF)n batteries within the US. During the discharge process, (CF) n was lithiated to generate LiF and C. Nevertheless, the reaction exhibited by the primary batteries made of lithium was not reversible, rendering them non-rechargeable during that period. In the 1970s, Sanyo, a leading manufacturer of lithium-ion batteries, developed the Li/MnO_2 battery. This innovative battery was quickly utilized in Sanyo's rechargeable electronic calculator, which was operated by solar energy. General Motors (GM) and Argonne National Laboratory (ANL) began their research by examining lithium-ion battery systems that utilize molten salts as electrolytes (Choi & Wang, 2018).

The electrodes in melting salt electrolyte batteries are constructed using molten lithium and sulfur sources that serve as the raw materials for their production. The electrolyte utilizes a combination of lithium salt melt, specifically LiCl and KCl. On the other hand, this particular type of battery necessitates a rather extreme temperature of 45 Celsius. Unfortunately, this particular battery fails to mitigate the environmental impact during cycling, resulting in a rapid deterioration of its capacity. In a recent study, the researchers made a significant shift in their research focus by exploring alternative materials for the anode in electric vehicle batteries. Instead of using lithium and sulfur, it turned its attention to iron sulfides (FeS and FeS_2) and lithium aluminum alloys (LiAl). This change resulted in notable improvements in the battery's cycle performance. Thanks to the quick progress of organic electrolyte battery technologies, the research on elevated temperatures of molten salt lithium-ion batteries in the 1990s came to a halt (Figure 5.4) (Wakihara, 2001).

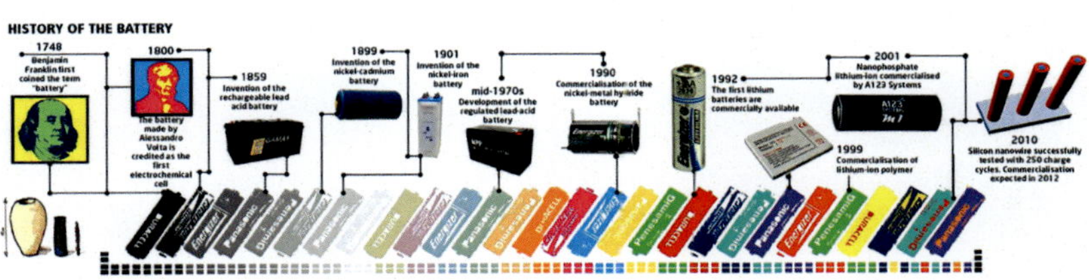

Figure 5.4. *Illustration of development of battery systems.*

Source: Haitao Song; Creative Commons License.

In 1976, Whittingham from Exxon Company introduced the groundbreaking lithium-ion battery. The lithium-ion battery utilizes lithium-ion compounds as a source of lithium, eliminating the need for elemental lithium within the battery. Whittingham made a groundbreaking discovery by demonstrating that layered can undergo electrochemical reactions via metallic lithium at room temperature. This breakthrough also marked the first instance of combining organic insertion chemical reactions with energy storage. The Li/ battery operates at approximately 2 V and exhibits exceptional cycle performance. The capacity degradation is minimal, with only a 0.05% decrease per cycle throughout 1,000 cycles. Exxon Company made a significant innovation by using LiAl alloy to be the negative electrode for the lithium-ion battery. This groundbreaking technology first came to consumers in 1977 and quickly found its way into watches and smaller electronic gadgets. Throughout the years, a range of compounds have been employed as materials for anode for lithium-ion batteries. These include VSe_2, MoO_3, V_6O_3, $CuTi_2S_4$, V_2O_5, and LiV_3O_8. These compounds can deintercalate with lithium reversibly. In 1980, Moli Energies of Canada successfully developed a new product by chemically combining lithium and MoS_2. This breakthrough greatly enhanced the electrical cycle efficiency of MoS_2 and paved the way for its industrialization.

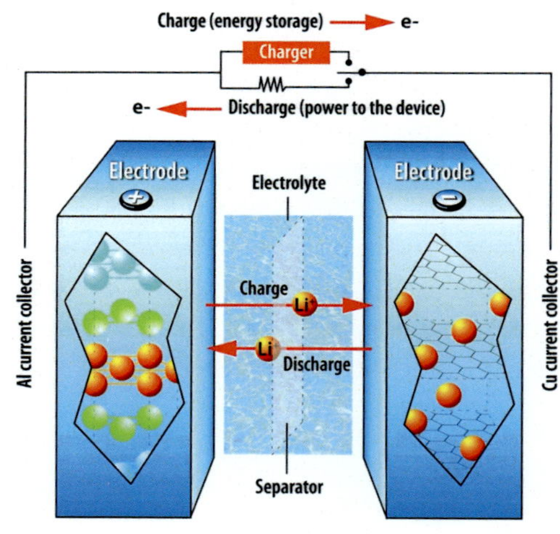

Figure 5.5. *Illustration of charge–discharge mechanism of Li-ion.*

Source: Hongliang; Creative Commons License.

This represents a significant step towards the development of modern lithium-ion batteries. Significant progress has been achieved in the study of lithium-ion batteries since the 1980s. In 1980, a breakthrough occurred when $LiCoO_2$ was introduced as a cathode by John Goodenough's group. In 1981, Bell Labs made a significant breakthrough by utilizing graphite to be the anode substance for batteries made of lithium-ion. In that year, a breakthrough was made by Japanese scientist Yoshino Akira. He found that conductive polyacetylene had the potential to serve as the negative electrode in the Charge-Discharge mechanisms of Li-ion rechargeable batteries, using organic solvents (Shen & Wang, 2019) (Figure 5.5).

In the realm of technological progress, a significant breakthrough occurred when Yoshino Akira successfully created an early version of a battery made of lithium-ion in 1983. This remarkable achievement involved the removal of metallic lithium from the battery, resulting in enhanced safety measures. On the other hand, the polyacetylene used as a negative electrode has certain drawbacks such as its low specific gravity, limited ability to enhance battery capacity, and inadequate stability. Yoshino Akira persisted in his efforts to refine and utilize materials made from carbon as a negative electrode, while still relying on lithium cobalt oxide for the positive electrode. In 1985, a remarkable milestone was achieved with the creation of a new lithium-ion secondary battery (LIB), that served as the basis for the development of modern lithium-ion batteries. Manthiram and Goodenough in 1989, made a significant discovery regarding the enhancement of working voltage in metal oxides through the induction impact of polyanions, specifically (such as SO_4^{2-}). In the year 1990, Sony's commercial operations lithium-ion batteries ($C/LiCoO_2$) achieved a breakthrough by transforming into a genuine lithium-ion battery (Aaldering & Song, 2019). A new type of lithium secondary battery has been developed, making use of graphitized carbon materials as the negative electrode. The battery's component framework includes a combination of different materials that work together to power the electric vehicle. In 1994, Tarascon & Guyomard made a breakthrough in the field of electrolyte systems by successfully utilizing ethylene carbonate and dimethyl carbonate. In 1997, a significant breakthrough was made when Goodenough discovered a new cathode material known as $LiFePO_4$. The lithium-ion battery systems had previously achieved full establishment (Zubi et al., 2018).

Practice Problem

Consider a hybrid electric vehicle (HEV) with a fuel efficiency of 50 miles per gallon (mpg) when running solely on its internal combustion engine. Additionally, this HEV is equipped with an electric motor capable of providing an extra 20 miles per kilowatt-hour (kWh). To determine the equivalent fuel efficiency in mpg when both the internal combustion engine and the electric motor are utilized, we need to account for the energy equivalence between gasoline and electricity. By considering that 1 gallon of gasoline is equal to 33.7 kilowatt-hours (kWh), we may determine the overall fuel efficiency in equivalent terms.

Solutions to Practice Problem

Firstly, we convert the additional mileage provided by the electric motor into gallons of gasoline equivalent. Given that an electric motor can travel 20 miles on one kWh of gasoline, one gallon of gasoline is equal to 33.7 kWh, it yields 20 miles * (1 kWh/33.7 kWh/gallon) ≈ 0.594 mpg. Adding this to the fuel efficiency of the internal combustion engine, the total equivalent fuel efficiency becomes approximately 50.594 mpg. This calculation accounts for the combined efficiency of both the electric motor and internal combustion, offering a comprehensive measure of the hybrid vehicle's fuel economy when utilizing both power sources.

At present, lithium-ion batteries are readily accessible as a form of secondary battery, known for their remarkable energy density and extended cycle life. Currently, there is a wide range of applications for advanced technology in areas such as mobile communication, electric automobiles, digital technologies, energy storage, and many other fields. There is a significant demand for lithium-ion batteries and their components in the future, with a corresponding market in the upstream and downstream production networks. Electric vehicles have seen significant growth thanks to the rapid advancements in lithium-ion battery technology. These include pure electric vehicles, hybrid vehicles, and extended-range vehicles. This growth is in line with the global expansion of the new energy vehicle industry, with a particular focus on pure electric vehicles. The lithium-ion battery is a widely adopted energy storage device with high energy density, making it a popular choice for powering various new energy vehicles. By the end of 2025, worldwide demand for lithium-ion batteries is projected to reach up to 200 million kWh in the electric vehicle market. Thus, apart from the quantity of battery production, having an exact grasp and command of power batteries has emerged as a crucial necessity for the progress of electric vehicles. Furthermore, a precise evaluation of the deterioration of power batteries in electric vehicles can contribute to the development of a more effective battery management system, resulting in enhanced battery performance and longevity (Armand et al., 2020).

Various research directions have categorized battery models into two main types: mechanism models and empirical models. There are different types of empirical models used in the field of electric vehicles. These models utilize various circuit and impedance models, which make use of electrical impedance spectroscopy. There are two main models for understanding the mechanism of electric vehicle batteries. An electrochemical model is utilized to study the chemical reactions occurring within the battery. The other is a simpler electrochemical model designed for easier practical application. Currently, the most efficient electric vehicle battery management techniques rely on battery experience models that are based on equivalent circuits. Due to its simple structure and ability to accurately simulate battery charging and discharging dynamics, the model of an equivalent circuit has gained significant popularity in recent engineering applications. Nevertheless, the battery model continues to face certain challenges when it comes to real-world implementation. Firstly, the electrochemical reaction system of the battery is often disregarded, along with the model parameters utilized lack a distinct electrochemical significance. As a result, it becomes challenging to accurately represent the actual characteristics of the battery. Additionally, while using the battery, other reactions occur alongside the electrochemical reactions inside the battery. The battery's efficiency can be affected by irreversible side reactions, leading to getting older and degradation. The precision of the battery model's calculation may be affected by this. Thus, battery electrochemical models encompass the internal electrochemical concept of the battery to establish a comprehensive array of mathematical models that establish a connection between fixed battery parameters for each component and battery parameters that vary with battery cycles. The electrochemical battery model has been found to accurately represent the battery's performance under different operating conditions during vehicle operation (Horiba, 2014).

5.2.2. The EV Charging Technology Development

The United States Department of Energy has classified EV charging into three levels: There are different levels of charging available for electric vehicles. Level 1 charging operates at an electrical power that is less than 5 kilowatts. Level 2 charging operates at a power range of 5 to 50 kW, whereas Level 3 charge is classified as super-fast and operates at a power level beyond 50 kW. Level 3 charging often utilizes direct current (DC) for power transmission, whereas Level 1 and Level 2 charging commonly employ onboard electronic converters to allow the transfer of alternating current (AC) energy. Rapid charging is essential for improving consumer convenience and making sure drivers are not limited by the typical daily driving range (Hemavathi & Shinisha, 2022). Additionally, the goal is to enhance the rate of energy transfer to be on par with traditional refueling stations for gasoline and diesel vehicles. It is important to consider that during the charging process, it is necessary to use a specific upper voltage limit to avoid overcharging the battery. It is widely understood that the charging current decreases as the state of charge increases (Ahmad et al., 2018).

In recent years, significant progress has been made in the development of high-density batteries, which has effectively mitigated the issue of "Range anxiety" that is frequently linked with EVs. With the increasing popularity of electric vehicles, a new issue known as "charging anxiety" is arising, replacing the previous concern regarding limited range. The EV charging system is essential for conditioning components and transferring energy within a stable network to charge the battery of an electric vehicle while it is connected. Reliable and effective charging methods are essential for a widespread transition to electric vehicles. These strategies might not only provide additional opportunities for drivers to generate income, but also increase awareness and instill confidence in prospective electric vehicle owners. Various techniques and strategies are employed to replenish the battery of electric vehicles. The process of battery charging involves the transmission of energy to the vehicle, while the charging strategy of the battery determines the particular curves of voltage and current that are utilized throughout the charging process (Chen et al., 2020).

In general, a charging strategy for lithium-ion batteries can be divided into three categories based on the underlying mathematical frameworks. There are different methodologies in the first category, including constant-current, constant-current (CC), multi-stage CCCV, constant-voltage (CCCV), and pulse charging techniques. These methods are recognized for their established charging profiles that feature, fixed voltage, power limits, and current, regardless of the battery's dynamic responses. Considering the practicality of model-free methods, the related programs are commonly regarded as a heuristic strategy. Thus, it is crucial to explore novel charging techniques that can meet the demand for fast charging while also mitigating any negative impact on battery longevity (Hans & Gupta, 2020).

Another way to categorize charging strategies is by using empirical models, like circuit-based models and neural network models. These models use historical experimental data to assess the states of batteries and determine electrical components. Several circuit instances, Kalman-type filters, recursive least squares, sliding mode observers, and moving horizon methods of estimation were utilized to accurately ascertain the battery states. Additionally, various methods were developed to enhance the charging performances, including linear quadratic control, frequency optimization, fuzzy control, multi-objective optimization, and model predictive control. The empirical calculations are efficient and straightforward but do not account for the physical aspects and battery deterioration. Thus, a charging control protocol based on empirical models may experience issues after a certain number of cycles.

Charging algorithms that rely on electrochemical models are quite complex. These models are controlled by kinetics and transport equations and can be used to solve optimization problems that aim to reduce charging time and account for uncertainties and disturbances in the battery model. Furthermore, these algorithms are capable of predicting temperature variations using thermal-related relations. As a result, the electrochemistry-based regulation algorithm serves as an accurate state observer for the battery mechanism. However, the intricate computational obstacles created by full-order nonlinear partial differential equations (PDEs) impede the feasible integration of a charging controller that operates in real time (Guo & Chan, 2015).

5.2.2.1. Constant-Current Constant Voltage-Related Charging Methods

The CC/CV charging algorithms are widely employed for charging lithium-ion batteries due to its straightforwardness and user-friendliness. The battery is charged utilizing the CC/CV algorithm, which involves applying a continuous current until the battery voltage reaches its maximum charging voltage. Once the voltage for charging is reached, it remains constant until the current decreases to a minimum level. Figure 5.6 illustrates the voltage curve of the CC/CV (Shafiei et al., 2018).

Several different versions of the CC/CV charging method were created. The charge profile is segmented into three distinct stages, as depicted in Figure 5.8, specifically referred to as CC-I, CC-II, and CV-I. The algorithm depends on the gradual increase in internal resistance during the charging process. Thanks to advancements in technology, electric vehicles now can apply the highest current because of the cell's smaller resistance to the lower SOC range. In the final two stages, the cell experiences a rapid increase in internal resistance. The experimental results confirm the effectiveness of the procedure in preventing significant temperature increases and prolonging the lifespan of the system (Gonzalez & Becerra, 2024).

A different approach to the multiple phases CC/CV method is shown in Figure 5.8. A high maximum voltage is utilized to rapidly charge the battery during the initial constant voltage (CV) mode, followed by a smooth transition to the conventional constant current-constant voltage (CCCV) method. This charging technique allows for the complete recharge of a fully discharged battery without causing any additional degradation (Gao et al., 2019) (Figures 5.6–5.8).

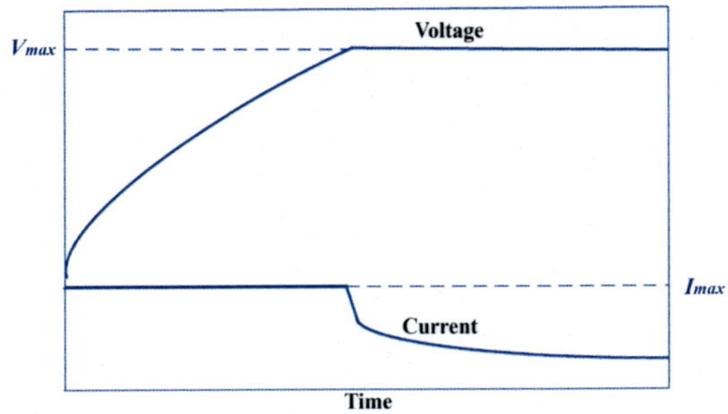

Figure 5.6. *Illustration of constant current–constant voltage curve.*

Source: Noaki Shinora; Creative Commons License.

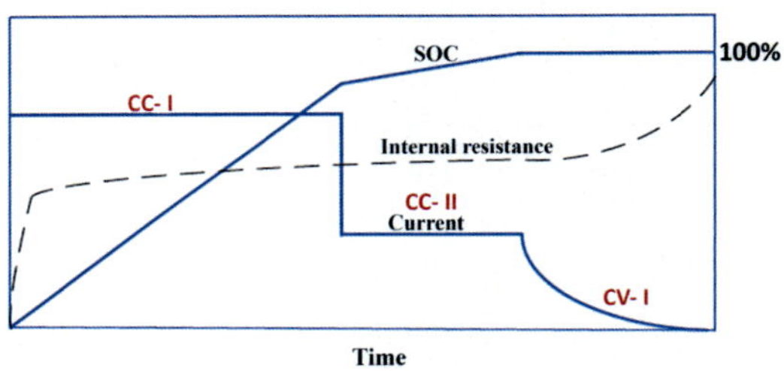

Figure 5.7. *Illustration of the present condition, level of electrical charge, and resistance patterns of the rapid charging method.*

Source: Noaki Shinora; Creative Commons License.

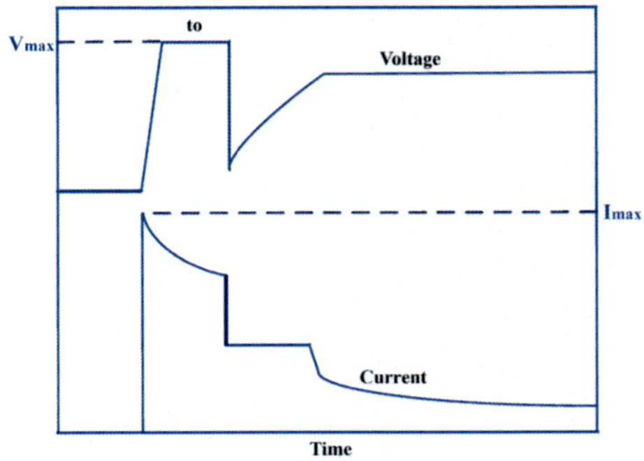

Figure 5.8. *Illustration of voltage and current characteristics of the boost-charging technique.*

Source: Balamurugan, S.; Creative Commons License.

5.2.2.2. Pulse Charge-Based Charging Curve

There is a charging algorithm called pulse charge that has been hailed for its speed and efficiency in charging lithium-ion batteries. The pulse charging method efficiently charges the battery by carefully managing factors such as battery heating, polarization, state of charge (SOC), and variable battery impedance. This technique is illustrated in Figure 5.9. During the pulse period, there is a designated rest time that enables the ions to diffuse within the electrode materials. This results in a more efficient charging process. Advanced charging techniques have been developed to improve the effectiveness of charging converters and are applied to lithium-ion batteries. These techniques involve applying small discharges to a battery through the rest period of pulse charging. The decrease in negative impulse has a stress-reducing effect on the cell, which helps to minimize any increase in temperature. With the constant push for innovation, engineers have developed circuit configurations that efficiently recapture the energy lost during the negative pulse, resulting in improved battery performance. Through the periodic depolarization of the cell, it becomes possible to consistently inject high currents into the battery (Jiang et al., 2021). This allows for a faster charging rate and reduced charging time. This approach enhances the chemical reactions occurring within the battery, leading to a substantial enhancement in battery lifespan (Li et al., 2019).

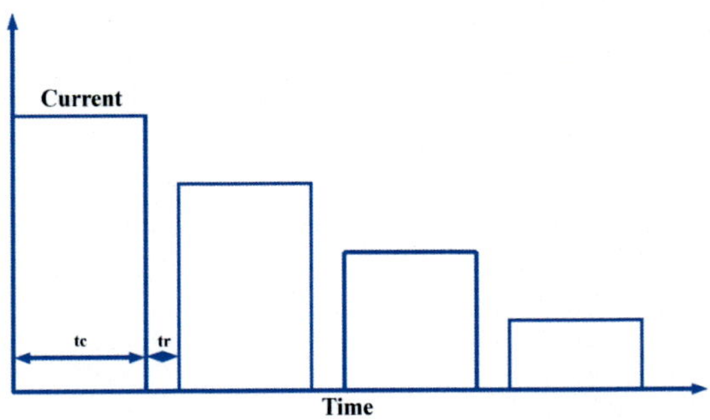

Figure 5.9. *Illustration of technique for lowering pulse charge in a nonlinear manner.*

Source: Albert Rayan; Creative Commons License.

5.2.2.3. Generalized Active Optimal Charging Algorithm

TAn active battery power management system typically has three essential components. The three crucial elements in this domain include the state estimator, model-based controller, and battery model. An ideal charging approach for the battery was achieved by implementing a closed-loop control system that incorporates process noise (Cao & Chen, 2018). This control structure is illustrated in Figure 5.10.

Did You Know?

There are various charging modes available for electric vehicles, each with its advantages in terms of speed, convenience, and compatibility with current infrastructure. These modes involve conductive charging, inductive charging, and mechanical battery swapping.

The battery model had been carefully designed to be straightforward and efficient, enabling precise control. The objective of developing this battery model is to accurately simulate the behavior of an actual battery system, taking into account a specific current profile (Deng et al., 2018). The output variables of the model, when paired with a noise vector, are used as input for the observer. Considering the complexity of the battery, it includes numerous state variables that are difficult to measure, including concentration and over-potential. Therefore, it is crucial to have a reliable and efficient model-based estimator to monitor the internal state of the batteries. A novel charging method has been developed, employing an advanced control algorithm and a simplified battery model with a state estimator. Additionally, when aiming for improved charging performance, it is crucial to take into account various factors related to battery health, such as temperature rise and side reaction rate, as indicators for desired outcomes (Gan et al., 2012).

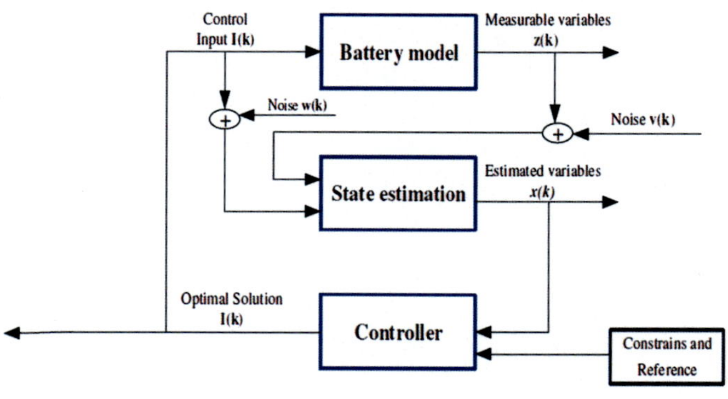

Figure 5.10. *Illustration of efficient charging infrastructure.*

Source: Hongliang; Creative Commons License.

When it comes to charging targets, it's important to consider two key factors: temperature rise and aging rate, both of which directly impact battery performance. Therefore, a charging method is formulated that considers the performance of the battery and the speed of charging concurrently (Zhang et al., 2018).

The properties associated with battery aging are selected from internal states. Undesirable outcomes often arise from side reactions implementing cathode and anode electrodes, along with the formation of the solid electrolyte interface (SEI). The concentration of the electrolyte is also considered a factor that can contribute to potential side effects. When it comes to the mechanical stress caused by diffusion, it is important to note that the lithium concentration on the surface of particles has a minimum limit of zero (Wu et al., 2022). By utilizing empirical equations that establish a connection between capacity loss and overall discharge Ah, which is the state of health (SOH) selected as a metric to assess degradation. There is a possibility that the rise in temperature may lead to battery thermal runaway. Therefore, battery temperature is regulated in conjunction with measures to prevent aging in the majority of instances. To address the optimal charging issue, it is crucial to develop a practical charging strategy that takes into account the characteristics of the battery (Liu et al., 2017).

5.3. PROSPECTS OF WIRELESS POWER TRANSFER TECHNOLOGIES

Learning Objectives:

- Evaluate the potential of wireless power transfer (WPT) for EV charging infrastructure.
- Analyze the integration of WPT with automated parking and driving.

Through extensive research, the WPT technology for EVs shows great potential for development. By incorporating WPT technologies into the EV's charging infrastructure, the next generation of EVs will be able to utilize a contactless charging method. This advancement holds significant significance in driving the widespread adoption and industrial implementation of new energy vehicles (Huang et al., 2019).

5.3.1. The Multi-Layer Wireless Charging Parking Lot

A patented conceptual electric vehicle charging station known as the Charging Tower has been developed by Ennead Lab, which is specifically designed for urban development. This urban roadside beacon utilizes wireless charging and stack-parking technologies to accommodate a high volume of electric cars while maintaining pedestrian-friendliness (Cheng et al., 2020).

In contrast to conventional gas stations, EVs are known for their eco-friendly nature. However, one drawback of EVs is their longer charging time, often exceeding five hours, which can be inconvenient for car owners who spend time waiting. Within such conditions, the Charging Tower comes into existence. The Charging Tower is structured with nine floors, enabling EVs to be parked within a multi-layer stacking configuration, resulting in significantly improved efficiency. The innovative design of the Charging Tower maximizes space utilization in urban areas with high population density, enabling the simultaneous charging of a larger number of electric vehicles within a compact footprint (Iqbal et al., 2019). Consequently, the amount of time car owners have to wait can be significantly decreased. The Charging Tower offers the convenience of charging multiple electric vehicles simultaneously, while also providing car owners with the flexibility to choose between standard charging and supercharging based on their individual preferences. Supercharging offers significantly faster charging speeds compared to standard charging. Additionally, supercharging stations are conveniently located at ground level, in contrast to standard charging points which are typically situated at higher levels (Qin & Costinett, 2021).

In the future, it is anticipated that Charging Towers will supplant current charging stations and gas stations, streamlining the charging procedure by allowing EVs to be charged conveniently while parked on the side of the road (Gunasekaran et al., 2017). Thanks to the Charging Tower's two charging options, car owners who choose standard charging can make the most of their charging time by exploring different locations and having a great time. For individuals who choose to enhance their vehicles, the Charging Tower serves as a conventional refueling station, allowing them to swiftly resume their journey following a rapid recharge (Balfaqih et al., 2021).

5.3.2. Driving and Automatic Parking Based on Wireless Charging

Currently, the process of charging an electric car involves connecting a battery gun to the electric vehicle. However, a significant number of charging stations currently operate without proper oversight. One intriguing question arises: could an electric vehicle autonomously full its charging process in the absence of occupants? Wireless charging offers a viable solution to the challenge of unmanned vehicles along with unattended charging stations. Wireless charging is a crucial technology that forms the basis for smart electric vehicles (Gandhi et al., 2021).

Without a doubt, the WPT system streamlines the charging procedure of EVs. When it comes to charging, one convenient option is to park the car inside a garage and let it get the power from the main side underneath. One potential solution is dynamic charging, which allows for charging the car while it is in motion, addressing any potential concerns regarding a range of electric vehicles. The time of automatic driving or intelligent cars is quickly approaching, and the significance of wireless charging in electric vehicles may not be emphasized enough (Gubbi et al., 2013). By integrating WPT and automated driving technologies, a seamless connection can be established between unmanned vehicles and unattended charging stations. This enables automatic and prolonged charging of the vehicles. In addition, the integration of various technologies enables the automatic allocation of charging space to optimize the utilization of charging ports. This not only enhances efficiency but also maximizes revenue by ensuring that the charging infrastructure is fully utilized (Rozman et al., 2019).

The integration of automated parking and driving capabilities with wireless charging technology can provide a seamless and convenient experience for electric vehicle owners. Nevertheless, achieving a fully autonomous electric vehicle is a future objective that necessitates the development of WPT technologies to facilitate its realization (Azad et al., 2018).

5.3.3. Automatic Access to an Energy Internet

With the increasing popularity of EVs, the concept of wireless charging is gaining traction as a convenient way to connect EVs to the energy Internet. Through the implementation of bidirectional flow technology, the possibility of automatically bidirectional portable charging and discharge for electric vehicles can be achieved. The integration of advanced technology enables a smooth transfer of energy among the power grid and electric vehicles, effectively turning the vehicles into self-sufficient and mobile energy storage units (Zeinab & Elmustafa, 2017). Simultaneously, within the context of the new energy micro-grid system, the utilization of wireless charging and discharging technology enables the secure and organized process of charging and discharging electric vehicles, while also facilitating their automatic involvement in the regulation of micro-grid energy. By promoting economic charging and grid coordination, electric vehicles are integrated into the latest energy interactive management system, playing a crucial role (Xie et al., 2022).

Through the integration of WPT technology and a fresh energy micro-network system, electric vehicles (EVs) have the capability to autonomously store electricity during periods of low demand and sell electricity through high demand charging periods. This is accomplished through the intelligent management and upkeep of multiple microsensors. Consequently, owners can benefit from the disparity in prices. The entire procedure is seamlessly executed through intelligent system control, eliminating the need for human intervention. This achievement can be seen as EVs achieving a level of self-sufficiency (Qiu & Gooi, 2000).

The development of the Intelligent Connected Vehicle (ICV) has evolved significantly due to the advancements made in artificial intelligence technology and improved sensor control technologies. The development of ICV encompasses two scientific pathways: networked connectivity and autonomous intelligence, ultimately culminating in the fusion of both (Wahla et al., 2019). The U.S. Department of Transportation highlighted the importance of connected vehicles (Connected Vehicles) and autonomous intelligent vehicles (Autonomous Vehicles) to be the main methods to achieve autonomous driving, as shown in Figure 5.11, in their "ITS Strategic Planning 2015–2019" report (Kurniawan, 2020).

The Internet of Vehicles (IOV) relies on a local area network (LAN), vehicle inter-network, and onboard internet (5G, etc.) by established communication protocols and data interaction standards. IOV facilitates communication and information exchange among various entities such as vehicles, roads, pedestrians, and the Internet. This enables dynamic information services, intelligent traffic management, and vehicle control (Kabalci et al., 2019).

The connected vehicle industry offers a wide range of information services via telematics and various types of connection and data interaction such as V2V, V2I, V2P, and V2X. This emerging industry represents a profound fusion of the automotive, and road transportation sectors and electronic information communication (Miah et al., 2019).

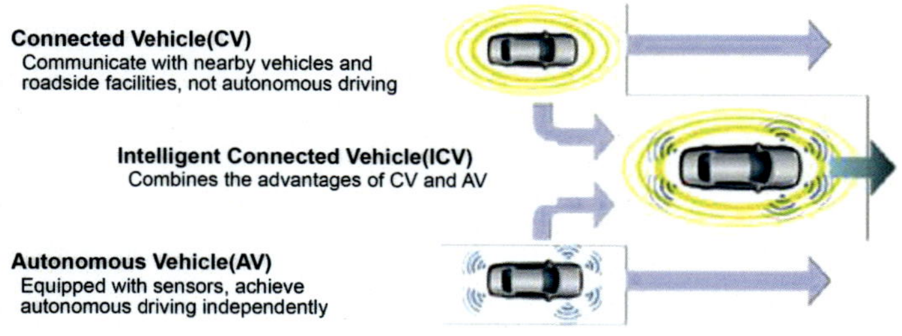

Figure 5.11. *Illustration of three technical paths towards achieving intelligent vehicles.*

Source: Ki Young Kim; Creative Commons License.

Modern vehicles are now equipped with advanced technology that allows them to operate autonomously. These vehicles utilize cutting-edge sensors, controllers, and actuators, and can seamlessly integrate principles from various fields such as automatic control, communication, artificial intelligence, and information networks. Intelligent information exchange is being implemented to tackle the difficulties of coordinating control and executing intelligent decision-making within complex environment perception. Thanks to recent advancements, driving has become safer, more comfortable, and highly efficient, potentially leading to a future where human drivers are no longer necessary (Effendy et al., 2021).

SAE categorizes the levels of intelligence into six distinct categories. As the level increases, so does the level of automation. Progressing from minimal intervention to complete autonomy, the levels of automation in vehicle technology are categorized as follows: partial automation, no automation, driving assistance, conditional automation, and full automation (BS, 2015). Autonomous vehicles (AVs) can be broken down into three primary elements: decision-making and planning, perception and location, and execution and control. These technical aspects are crucial for the functioning of AVs. These components work together within an integrated system to enable AV functionality (Ioannou et al., 2021).

SUMMARY

- The chapter "Technological Advances in Electric Vehicles" delves into the multifaceted landscape of electric vehicle (EV) technology, offering a comprehensive overview of its evolution and current state. It begins by tracing the historical trajectory of electric vehicles, highlighting key milestones and advancements that have shaped the industry.
- The discussion extends to the present, examining the development status of electric vehicles both domestically and abroad, providing insights into the global landscape of EV adoption and innovation.
- A significant portion of the chapter is dedicated to exploring charging technologies, including the emergence of wireless charging solutions for electric vehicles. It delves into both static and dynamic wireless charging methods, offering insights into their feasibility and potential impact on the EV ecosystem. Moreover, the chapter delves into the advancements in lithium-ion battery technology, which serves as the cornerstone of electric vehicle propulsion systems and discusses various charging technologies that complement these advancements.
- Looking towards the future, the chapter describes the prospects of wireless power transfer technologies for electric vehicles, envisioning scenarios where multi-layer wireless charging infrastructure could revolutionize the way EVs are charged and managed.
- Additionally, it explores the concept of automatic access to an energy internet, highlighting the potential for seamless integration of electric vehicles into a broader energy ecosystem.

REVIEW QUESTIONS

1. Summarize electric vehicle (EV) historical milestones and their impact on the industry.
2. Compare the advantages and limitations of BEVs and HEVs.
3. Discuss global EV development, adoption rates, and influencing factors.
4. Explain wireless EV charging, static vs. dynamic methods, and associated pros and cons.
5. Describe advancements in lithium-ion batteries and their impact on EVs.

6. Explore existing EV charging modes, their differences, and characteristics.
7. Analyze the potential of wireless power transfer for EVs, including multi-layer charging.
8. Investigate EV integration with the energy internet and its implications.
9. Evaluate the benefits and challenges of automatic parking and driving with wireless charging.

MULTIPLE CHOICE QUESTIONS

1. **Which type of electric vehicle relies solely on battery power for propulsion?**
    a. Hybrid Electric Vehicle (HEV)
    b. Plug-in Hybrid Electric Vehicle (PHEV)
    c. Battery Electric Vehicle (BEV)
    d. Fuel Cell Electric Vehicle (FCEV)
2. **What is the primary function of dynamic wireless charging for electric vehicles?**
    a. Charging the vehicle while it is stationary
    b. Charging the vehicle while it is in motion
    c. Charging the vehicle using solar panels
    d. Charging the vehicle using kinetic energy recovery
3. **Which component serves as the cornerstone of electric vehicle propulsion systems?**
    a. Lithium-ion Battery
    b. Electric Motor
    c. Fuel Cell
    d. Internal Combustion Engine
4. **What is the focus of automatic access mechanisms for electric vehicles to an energy internet?**
    a. Regulating vehicle speed based on energy availability
    b. Automatically connecting EVs to the internet for entertainment purposes
    c. Integrating EVs into a broader energy ecosystem
    d. Enabling remote control of EV charging stations

5. **What technology enables electric vehicles to charge without physical contact with a charging station?**
   a. Dynamic Charging
   b. Inductive Charging
   c. Supercharging
   d. Solar Charging
6. **What technology allows electric vehicles to automatically park and charge wirelessly?**
   a. Automatic Parking Assist
   b. Dynamic Wireless Charging
   c. Multi-Layer Wireless Charging
   d. Automatic Energy Access Protocol

Answers to Multiple Questions

1. (c); 2. (b); 3. (a); 4. (c); 5. (b) 6. (c)

REFERENCES

1. Aaldering, L. J., & Song, C. H. (2019). Tracing the technological development trajectory in post-lithium-ion battery technologies: A patent-based approach. *Journal of Cleaner Production, 241*, 118343.
2. Agrawal, M., & Rajapatel, M. S. (2020). Global perspective on electric vehicle 2020. *International Journal of Engineering Research & Technology, 9*(1), 8–11.
3. Ahmad, A., Alam, M. S., & Chabaan, R. (2017). A comprehensive review of wireless charging technologies for electric vehicles. *IEEE Transactions on Transportation Electrification, 4*(1), 38–63.
4. Ahmad, A., Khan, Z. A., Saad Alam, M., & Khateeb, S. (2018). A review of the electric vehicle charging techniques, standards, progression and evolution of EV technologies in Germany. *Smart Science, 6*(1), 36–53.
5. Ajanovic, A. (2015). The future of electric vehicles: Prospects and impediments. *Wiley Interdisciplinary Reviews: Energy and Environment, 4*(6), 521–536.
6. Amjad, M., Farooq-i-Azam, M., Ni, Q., Dong, M., & Ansari, E. A. (2022). Wireless charging systems for electric vehicles. *Renewable and Sustainable Energy Reviews, 167*, 112730.
7. Aretxabaleta, I., De Alegria, I. M., Andreu, J., Kortabarria, I., & Robles,

E. (2021). High-voltage stations for electric vehicle fast-charging: Trends, standards, charging modes and comparison of unity power-factor rectifiers. *IEEE Access, 9*, 102177–102194.

8. Armand, M., Axmann, P., Bresser, D., Copley, M., Edström, K., Ekberg, C., & Zhang, H. (2020). Lithium-ion batteries–Current state of the art and anticipated developments. *Journal of Power Sources, 479*, 228708.
9. Asif, U., & Schmidt, K. (2021). Fuel cell electric vehicles (FCEV): Policy advances to enhance commercial success. *Sustainability, 13*(9), 5149.
10. Azad, A. N., Echols, A., Kulyukin, V. A., Zane, R., & Pantic, Z. (2018). Analysis, optimization, and demonstration of a vehicular detection system intended for dynamic wireless charging applications. *IEEE Transactions on Transportation Electrification, 5*(1), 147–161.
11. Balfaqih, M., Jabbar, W., Khayyat, M., & Hassan, R. (2021). Design and development of smart parking system based on fog computing and internet of things. *Electronics, 10*(24), 3184.
12. Barkenbus, J. N. (2020). Prospects for electric vehicles. *Sustainability, 12*(14), 5813.
13. BS, R. (2015). Automatic smart parking system using Internet of Things (IOT). *International Journal of Scientific and Research Publications, 5*(12), 629–632.
14. Cairns, E. J., & Albertus, P. (2010). Batteries for electric and hybrid-electric vehicles. *Annual Review of Chemical and Biomolecular Engineering, 1*, 299–320.
15. Cao, C., & Chen, B. (2018). Generalized Nash equilibrium problem-based electric vehicle charging management in distribution networks. *International Journal of Energy Research, 42*(15), 4584–4596.
16. Chen, T., Zhang, X. P., Wang, J., Li, J., Wu, C., Hu, M., & Bian, H. (2020). A review on electric vehicle charging infrastructure development in the UK. *Journal of Modern Power Systems and Clean Energy, 8*(2), 193–205.
17. Chen, X., Zhang, H., Xu, Z., Nielsen, C. P., McElroy, M. B., & Lv, J. (2018). Impacts of fleet types and charging modes for electric vehicles on emissions under different penetrations of wind power. *Nature Energy, 3*(5), 413–421.
18. Cheng, B., Lu, J., Zhang, Y., Pan, G., Chabaan, R., & Mi, C. C. (2020). A metal object detection system with multilayer detection coil layouts for electric vehicle wireless charging. *Energies, 13*(11), 2960.
19. Choi, S., & Wang, G. (2018). Advanced lithium-ion batteries for practical applications: Technology, development, and future perspectives. *Advanced Materials Technologies, 3*(9), 1700376.

20. Climent, H., Pla, B., Bares, P., & Pandey, V. (2021). Exploiting driving history for optimizing the Energy Management in plug-in Hybrid Electric Vehicles. *Energy Conversion and Management, 234*, 113919.
21. Cuma, M. U., & Koroglu, T. (2015). A comprehensive review on estimation strategies used in hybrid and battery electric vehicles. *Renewable and Sustainable Energy Reviews, 42*, 517–531.
22. Deng, Z., Liu, M., Chen, H., Lu, W., & Dong, P. (2018). Optimal scheduling of active distribution networks with limited switching operations using mixed-integer dynamic optimization. *IEEE Transactions on Smart Grid, 10*(4), 4221–4234.
23. Dericioglu, C., YiriK, E., Unal, E., Cuma, M. U., Onur, B., & Tumay, M. (2018). A review of charging technologies for commercial electric vehicles. *International Journal of Advances on Automotive and Technology, 2*(1), 61–70.
24. Du, J., & Ouyang, D. (2017). Progress of Chinese electric vehicles industrialization in 2015: A review. *Applied Energy, 188*, 529–546.
25. Effendy, M. E. S., Kusumanto, R. D., & Carlos, R. S. (2021). Automatic Control System for Hybrid Electrical Energy Supply Based on Internet of Things. *International Journal of Research in Vocational Studies (IJRVOCAS), 1*(2), 34–44.
26. Galus, M. D., Zima, M., & Andersson, G. (2010). On integration of plug-in hybrid electric vehicles into existing power system structures. *Energy Policy, 38*(11), 6736–6745.
27. Gan, L., Topcu, U., & Low, S. H. (2012). Optimal decentralized protocol for electric vehicle charging. *IEEE Transactions on Power Systems, 28*(2), 940–951.
28. Gandhi, R., Nagarajan, S., Chandramohan, J., Parimala, A., & Arulmurugan, V. S. (2021). IoT based automatic smart parking system with EV-charging point in crowd sensing area. *Annals of the Romanian Society for Cell Biology, 25*(6), 6398–6409.
29. Gao, Y., Zhang, X., Cheng, Q., Guo, B., & Yang, J. (2019). Classification and review of the charging strategies for commercial lithium-ion batteries. *IEEE Access, 7*, 43511–43524.
30. Gönül, Ö., Duman, A. C., & Güler, Ö. (2021). Electric vehicles and charging infrastructure in Turkey: An overview. *Renewable and Sustainable Energy Reviews, 143*, 110913.
31. Gonzalez-Saenz, J., & Becerra, V. (2024). Determining Fast Battery Charging Profiles Using an Equivalent Circuit Model and a Direct Optimal Control Approach. *Energies, 17*(6), 1470.

32. Gubbi, J., Buyya, R., Marusic, S., & Palaniswami, M. (2013). Internet of Things (IoT): A vision, architectural elements, and future directions. *Future Generation Computer Systems, 29*(7), 1645–1660.

33. Gunasekaran, V., Park, G. H., Suemitsu, M., & Fukidome, H. (2017). Fabrication of multi-layer Bi2Se3 devices and observation of anomalous electrical transport behaviors. *Materials Science in Semiconductor Processing, 68*, 128–132.

34. Guo, C., & Chan, C. C. (2015). Analysis method and utilization mechanism of the overall value of EV charging. *Energy Conversion and Management, 89*, 420–426.

35. Hans, N., & Gupta, M. S. (2020). Trends In Electric Vehicle (EV) Charging and Key Technology Developments. *International Journal of Engineering Research and, 9*(09), 44–48.

36. Hemavathi, S., & Shinisha, A. (2022). A study on trends and developments in electric vehicle charging technologies. *Journal of Energy Storage, 52*, 105013.

37. Horiba, T. (2014). Lithium-ion battery systems. *Proceedings of the IEEE, 102*(6), 939–950.

38. Hu, Y., Song, X., Cao, W., & Ji, B. (2014). New SR drive with integrated charging capacity for plug-in hybrid electric vehicles (PHEVs). *IEEE Transactions on Industrial Electronics, 61*(10), 5722–5731.

39. Huang, J., Zhou, Y., Ning, Z., & Gharavi, H. (2019). Wireless power transfer and energy harvesting: Current status and future prospects. *IEEE Wireless Communications, 26*(4), 163–169.

40. Ioannou, K., Karampatzakis, D., Amanatidis, P., Aggelopoulos, V., & Karmiris, I. (2021). Low-cost automatic weather stations in the internet of things. *Information, 12*(4), 146.

41. Iqbal, R., Maniak, T., & Karyotis, C. (2019). Intelligent remote monitoring of parking spaces using licensed and unlicensed wireless technologies. *IEEE Network, 33*(4), 23–29.

42. Jaiswal, M., & Zane, L. J. (2022). Drivers of sustainable new technology diffusion in national markets: The case of electric vehicles. *Thunderbird International Business Review, 64*(1), 25–38.

43. Jiang, L., Cao, Y., Li, Y., Zhao, Y., Liu, P., Huang, C., & Sun, Y. (2021). Optimal charging strategy with complementary pulse current control of lithium-ion battery for electric vehicles. *IEEE Transactions on Transportation Electrification, 8*(1), 62–71.

44. Kabalci, Y., Kabalci, E., Padmanaban, S., Holm-Nielsen, J. B., &

Blaabjerg, F. (2019). Internet of things applications as energy internet in smart grids and smart environments. *Electronics, 8*(9), 972.

45. Kubański, M. (2020). Prospects for the use of electric vehicles in public transport on the example of the city of Czechowice-Dziedzice. *Transportation Research Procedia, 44*, 110–114.
46. Kurniawan, N. (2020). Electrical energy monitoring system and Automatic Transfer Switch (ATS) Controller with the internet of things for solar power plants. *Journal of Soft Computing Exploration, 1*(1), 16–23.
47. Li, S., Wu, Q., Zhang, D., Liu, Z., He, Y., Wang, Z. L., & Sun, C. (2019). Effects of pulse charging on the performances of lithium-ion batteries. *Nano Energy, 56*, 555–562.
48. Liu, K., Li, K., & Zhang, C. (2017). Constrained generalized predictive control of battery charging process based on a coupled thermoelectric model. *Journal of Power Sources, 347*, 145–158.
49. Lukic, S., & Pantic, Z. (2013). Cutting the cord: Static and dynamic inductive wireless charging of electric vehicles. *IEEE Electrification Magazine, 1*(1), 57–64.
50. Mazumder, M., & Debbarma, S. (2020). EV charging stations with a provision of V2G and voltage support in a distribution network. *IEEE Systems Journal, 15*(1), 662–671.
51. Mehta, R., Srinivasan, D., Khambadkone, A. M., Yang, J., & Trivedi, A. (2016). Smart charging strategies for optimal integration of plug-in electric vehicles within existing distribution system infrastructure. *IEEE Transactions on Smart Grid, 9*(1), 299–312.
52. Miah, S., Islam, G. J., Das, S. K., Islam, S., Islam, M., & Islam, K. K. (2019). Internet of Things (IoT) based automatic electrical energy meter billing system. *IOSR Journal of Electronics and Communication Engineering, 14*(4(I)), 39–50.
53. Miller, J. M., Onar, O. C., White, C., Campbell, S., Coomer, C., Seiber, L., & Steyerl, A. (2014). Demonstrating dynamic wireless charging of an electric vehicle: The benefit of electrochemical capacitor smoothing. *IEEE Power Electronics Magazine, 1*(1), 12–24.
54. Mo, T., Li, Y., Lau, K. T., Poon, C. K., Wu, Y., & Luo, Y. (2022). Trends and emerging technologies for the development of electric vehicles. *Energies, 15*(17), 6271.
55. Mohamed, N., Aymen, F., Alharbi, T. E., El-Bayeh, C. Z., Lassaad, S., Ghoneim, S. S., & Eicker, U. (2022). A comprehensive analysis of wireless charging systems for electric vehicles. *IEEE Access, 10*, 43865–43881.

56. Ou, C. H., Liang, H., & Zhuang, W. (2014). Investigating wireless charging and mobility of electric vehicles on electricity market. *IEEE Transactions on Industrial Electronics, 62*(5), 3123–3133.
57. Panchal, C., Stegen, S., & Lu, J. (2018). Review of static and dynamic wireless electric vehicle charging system. *Engineering Science and Technology, an International Journal, 21*(5), 922–937.
58. Pettersson, S. (2015). Lightweight infrastructure for electric vehicle charging. *World Electric Vehicle Journal, 7*(4), 631–642.
59. Qin, R., Li, J., & Costinett, D. (2021). A 6.6-kW high-frequency wireless power transfer system for electric vehicle charging using multilayer nonuniform self-resonant coil at MHz. *IEEE Transactions on Power Electronics, 37*(4), 4842–4856.
60. Qiu, B., & Gooi, H. B. (2000). Web-based SCADA display systems (WSDS) for access via Internet. *IEEE Transactions on Power Systems, 15*(2), 681–686.
61. Rahman, S., Khan, I. A., Khan, A. A., Mallik, A., & Nadeem, M. F. (2022). Comprehensive review & impact analysis of integrating projected electric vehicle charging load to the existing low voltage distribution system. *Renewable and Sustainable Energy Reviews, 153*, 111756.
62. Rajashekara, K. (1994). History of electric vehicles in General Motors. *IEEE Transactions on Industry Applications, 30*(4), 897–904.
63. Rao, R., Zhang, X., Xie, J., & Ju, L. (2015). Optimizing electric vehicle users' charging behavior in battery swapping mode. *Applied Energy, 155*, 547–559.
64. Rozman, M., Ikpehai, A., Adebisi, B., Rabie, K. M., Gacanin, H., Ji, H., & Fernando, M. (2019). Smart wireless power transmission system for autonomous EV charging. *IEEE Access, 7*, 112240–112248.
65. Salkuti, S. R. (2023). Advanced technologies for energy storage and electric vehicles. *Energies, 16*(5), 2312.
66. Sanguesa, J. A., Torres-Sanz, V., Garrido, P., Martinez, F. J., & Marquez-Barja, J. M. (2021). A review on electric vehicles: Technologies and challenges. *Smart Cities, 4*(1), 372–404.
67. Shafiei, N., Ordonez, M., Tokaldani, M. A. S., & Arefifar, S. A. (2018). PV battery charger using an L3C resonant converter for electric vehicle applications. *IEEE Transactions on Transportation Electrification, 4*(1), 108–121.
68. Sharma, S., Panwar, A. K., & Tripathi, M. M. (2020). Storage technologies for electric vehicles. *Journal of Traffic and Transportation Engineering (English Edition), 7*(3), 340–361.

69. Shen, C., & Wang, H. (2019). Research on the technological development of lithium-ion battery industry in China. *Journal of Physics: Conference Series, 1347*(1), 012087.

70. Szumska, E. M. (2023). Electric vehicle charging infrastructure along highways in the EU. *Energies, 16*(2), 895.

71. Thomas, C. E. (2009). Fuel cell and battery electric vehicles compared. *International Journal of Hydrogen Energy, 34*(15), 6005–6020.

72. Wahla, A. H., Chen, L., Wang, Y., Chen, R., & Wu, F. (2019). Automatic wireless signal classification in multimedia Internet of Things: An adaptive boosting enabled approach. *IEEE Access, 7*, 160334–160344.

73. Wakihara, M. (2001). Recent developments in lithium-ion batteries. *Materials Science and Engineering: R: Reports, 33*(4), 109–134.

74. Wu, M., Su, L., Chen, J., Duan, X., Wu, D., Cheng, Y., & Jiang, Y. (2022). Development and prospect of wireless power transfer technology used to power unmanned aerial vehicle. *Electronics, 11*(15), 2297.

75. Xiao, Y., Liu, C., & Yu, F. (2018). An integrated on-board EV charger with safe charging operation for three-phase IPM motor. *IEEE Transactions on Industrial Electronics, 66*(10), 7551–7560.

76. Xie, L., Wu, J., Li, Y., Sun, Q., & Xi, L. (2022). Automatic generation control strategy for integrated energy system based on ubiquitous power internet of things. *IEEE Internet of Things Journal, 10*(9), 7645–7654.

77. Zeinab, K. A. M., & Elmustafa, S. A. A. (2017). Internet of things applications, challenges and related future technologies. *World Scientific News, 67*(2), 126–148.

78. Zhang, H., Moura, S. J., Hu, Z., Qi, W., & Song, Y. (2018). Joint PEV charging network and distributed PV generation planning based on accelerated generalized benders decomposition. *IEEE Transactions on Transportation Electrification, 4*(3), 789–803.

79. Zhang, X., Yuan, Z., Yang, Q., Li, Y., Zhu, J., & Li, Y. (2016). Coil design and efficiency analysis for dynamic wireless charging system for electric vehicles. *IEEE Transactions on Magnetics, 52*(7), 1–4.

80. Zubi, G., Dufo-López, R., Carvalho, M., & Pasaoglu, G. (2018). The lithium-ion battery: State of the art and future perspectives. *Renewable and Sustainable Energy Reviews, 89*, 292–308.

CHAPTER 6

INDUCTIVE WIRELESS POWER CHARGING FOR ELECTRIC VEHICLES

LEARNING OBJECTIVES

At the end of this chapter, readers will be able to:

1. Understand the principles and advantages of inductive power transfer (IPT) systems.
2. Explore the architecture and components of stationary inductive charging systems.
3. Know the challenges and opportunities associated with in-motion inductive charging for EVs.
4. Understand about compensation networks and their role in optimizing IPT efficiency.
5. Discover the current state of the art in IPT systems for electric vehicle charging.
6. Know the key challenges in standardization and infrastructure deployment for inductive charging.

INTRODUCTORY EXAMPLE

Qualcomm, an established leader in wireless technologies, is making significant strides in dynamic inductive charging for electric vehicles (EVs), thereby transforming the landscape of electric mobility. Qualcomm's advancements are having a significant impact on the sustainable transportation scene in busy urban centers such as downtown San Francisco. Electric automobiles that feature Qualcomm's technology move around city streets and effortlessly tap into the energy provided by dynamic charging pads incorporated beneath the asphalt. This advancement eliminates the necessity for frequent stops at conventional charging stations, enabling electric vehicles to increase their range and navigate effortlessly through urban surroundings.

Qualcomm's dynamic inductive charging technology showcases the company's commitment to advancing wireless power transfer. Qualcomm is leading the charge towards a more environmentally friendly future of transportation by developing technology that allows battery-powered cars to recharge while on the move. As cities around the globe embrace the shift towards electric mobility, Qualcomm's groundbreaking developments in dynamic charging hold great potential for influencing the infrastructure of urban transportation and decreasing dependence on fossil fuels.

Qualcomm's dynamic inductive charging technology is poised to revolutionize how electric vehicles are powered and propelled in downtown San Francisco and other areas. By seamlessly integrating charging infrastructure into urban streets, Qualcomm is enabling the advancement of electric vehicle technology and expediting the shift toward a more environmentally friendly transportation ecosystem. Thanks to Qualcomm's leadership, the prospect of electric mobility appears more promising than ever.

UNIT INTRODUCTION

In recent years, the public transport sector has become increasingly recognized as a major source of dangerous emissions that have negative impacts on both human health and the environment. The industry heavily depends on non-renewable petroleum products, which are projected to be depleted in the future. As a result, it is crucial to embrace electric vehicles (EVs) to be an attractive option to lessen our reliance on fossil fuels, diversify transportation fuel options, decrease consumption of energy in transportation due to improved EV efficiency, and ultimately reduce emissions and environmental harm. Nevertheless, the extensive implementation of electric vehicles encounters challenges like extended charging durations, concerns about limited range, and inadequate infrastructure preparedness (Mahesh et al., 2021). There are three main types of EV battery charging: conductive charging, battery swapping stations, and wireless charging. Wireless charging, in particular, enables the charging of EV batteries from a distance without the requirement of physical connections. There are three different techniques used for wireless charging: stationary charging, quasi-dynamic charging, and dynamic or in-motion charging. Wireless charging technology presents a range of advantages for electric vehicle (EV) charging, such as convenience, automatic functionality, and reliability even in harsh environments, as well as its resistance to damage. In addition, the removal of cables improves safety. However, similar to other technologies, wireless charging systems face certain limitations. These include efficiency loss caused by misalignment, high costs for units and installation, as well as complexities in design, infrastructure, manufacturing, and integration (Hutchinson et al., 2019).

The different types of WPT technologies can be categorized into four groups, which are far-field transfer, near-field transfer, mechanical interaction with permanent magnets (such as magnetic gear), and acoustic methods, as shown in Figure 6.1. Magnetic gear technology Wireless Power Transfer (WPT) was originally developed to replace traditional wired connections, but it has since been adapted for a wide range of uses, including stationary charging for electric vehicles, powering electric vehicles while in motion, and medical devices. The technology of Acoustic WPT involves a transmitter that transforms electric power into compressed waves of sound and a receiver that includes a transducer that transforms the motion generated by sound waves into electrical power (Lukic & Pantic, 2013).

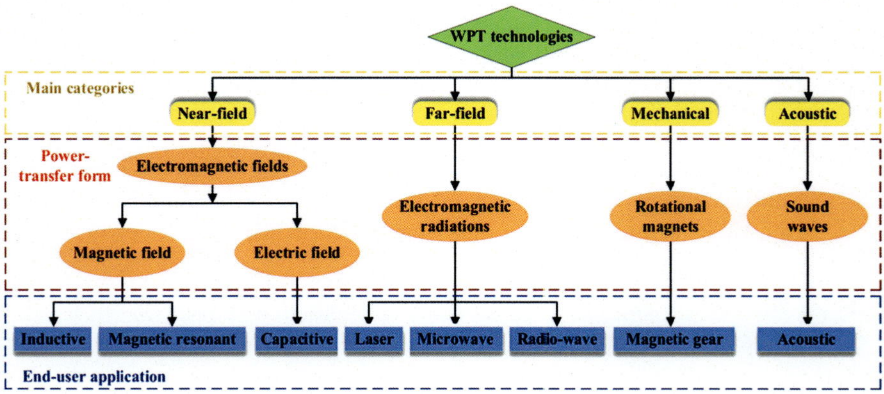

Figure 6.1. *Illustration of the classification of wireless power transfer technologies.*

Source: Shubhangi Das; Creative Commons License.

This unit delves into the complexities of inductive power transfer systems (IPT) along with their crucial role in the advancement of electric vehicle (EV) charging infrastructure. This unit provides a comprehensive examination of the technological advancements driving the evolution of wireless charging solutions. It explores topics that include magnetic coupler system architecture, stationary and in-motion IPT systems, and compensation networks. This unit offers valuable insights into the possibilities and hurdles that are forming the development of electric mobility. It examines the present-day state of the art in inductive power transfer (IPT) systems and addresses challenges related to stationary and dynamic charging for electric vehicles (EVs) (Buja et al., 2016).

6.1. INDUCTIVE POWER TRANSFER SYSTEMS

Learning Objectives:

- Understand the concept of Inductive power transfer system.

Inductive wireless power charging is a reliable and efficient technology enabling the transmission of power through the magnetic field generated through inductive coupling coils to transmit power through the air. It has a wide range of applications such as consumer electronics, biomedical devices, and electric vehicles (EVs) (Stielau et al., 2000). Inductive wireless power charging is considered highly suitable for electric vehicle (EV) charging because of multiple benefits, including:

- Ensuring the safety of humans by implementing electrical isolation for all components, resulting in a safer charging experience.
- Operating efficiently and reliably even in challenging environmental conditions like heavy rain, wind, and snow.
- An autonomous system that operates without any user involvement.
- The interoperability principle addresses concerns regarding compatibility among plugs and outlets.
- The ability to transfer the necessary power (ranging from a few kilowatts to megawatts) across significant air gaps of 100 to 400 millimeters, thus obliging the ground clearance of different electric vehicles (Covic & Boys, 2013).
- The lack of moving or rotating components leads to decreased noise levels and maintenance requirements.

A typical IPT charging system is comprised of two primary parts: the receiver pad, which is installed on the electric vehicle (EV), and the transmitter pad, which is concealed underground, as depicted in Figure 6.2. The transferring pad operates by high-frequency power generated through an HF inverter provided using rectified DC voltage (Mahmud et al., 2017). Compensation elements, such as inductors, capacitors, or an LC network, are strategically placed within the HF inverter and the transmitter pad to ensure the efficient transfer of nominal power. The coil on the transmitting pad generates an electromagnetic field that causes a high-frequency voltage and current to be produced in the coil on the receiving pad. The voltage is subsequently transmitted using a receiver compensation network and fixed to align with the battery's required voltage level (Madawala et al., 2011).

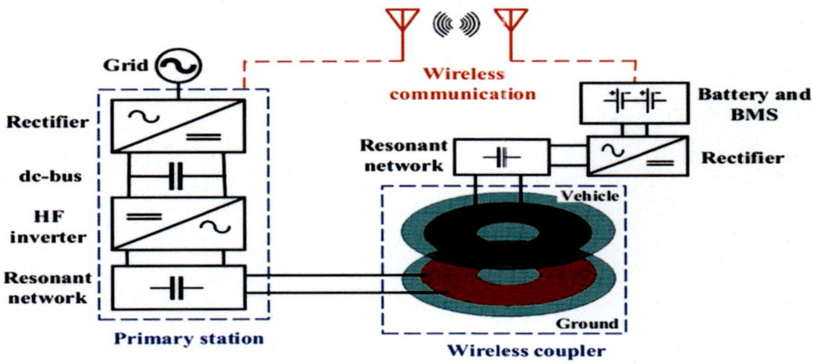

Figure 6.2. *Illustration of components of an exemplary inductive power transfer.*

Source: Philip, B.; Creative Commons License.

Did You Know?

The idea of inductive charging has its roots in the 19th century when physicist Michael Faraday made the groundbreaking discovery of electromagnetic induction.

According to the SAE J2954 standard, IPT systems usually operate at frequencies between 79 and 90 kHz, which allows for smaller transmitter and receiver sizes. There is a wireless communication link between the controllers of the receiver and the transmitter, which enables various services like payment processing, ensuring proper alignment, and notifying battery status (Kim et al., 2016).

6.2. MAGNETIC COUPLER SYSTEM ARCHITECTURE

Learning Objectives:

- Understand the essential components and configurations of stationary and in motion IPT charging systems.

The charging pad is an essential component of the IPT charging system as it enables a smooth transfer of electricity coming from the supply into the electric vehicle battery. The system is composed of three crucial components, each with its own set of requirements: conductive coils, magnetic cores that function as field concentrators, and EMF shielding. The combination of these elements ensures a smooth and efficient power transmission through the charging process, resulting in reliability and cost-effectiveness (Chakibanda & Komanapalli, 2023).

6.2.1. Stationary IPT

The configuration of the IPT pad is a critical factor for assessing the system's performance in terms of electromagnetic field safety, coupling factor, transmission efficiency, and tolerance for misalignment cases. In the past, U-cores, E-cores, and pot cores were commonly used in traditional transformers and were also explored for use in IPT systems (Hsieh et al., 2016). Nevertheless, these architectures present significant obstacles for stationary EV charging as a result of their considerable size, inflexible configuration, substantial weight, exorbitant cost, and restricted tolerance for lateral misalignment. References have suggested the utilization of planar coils for overcoming these obstacles. These designs focus on optimizing the three components (coil, core, and shield) to minimize the size, weight, and thickness of the power transmission pad. Additionally, they aim to improve its ability to handle misalignment, whether it's lateral or rotational (Mohamed et al., 2020).

Did You Know?

The progress in inductive charging technological advances is leading to the seamless integration of electric vehicles with smart grids, allowing for improved energy management and distribution.

There are three categories in which the charging pad architecture might be classified, depending on the nature and shape of its EMF components. Charging pads generate magnetic fields with different components, including circular and rectangular charging pads, as well as polarized charging pads, which generate horizontal magnetic field components, like DD and DDQ, primarily produce horizontal magnetic field components, third consists of charging pads with multiple coils, such as bipolar and tri-polar charging pads, which generate both perpendicular and horizontal field components (Thai et al., 2019).

Although the DDQ charging pad architecture and bipolar charging pad architecture are well-suited for car-side transmitters, other pad architectures, including DD, multi-coil homogeneous, circular, and rectangular pad architectures, are generally more effective on the transmitter side. In addition, many of these kinds of architectures successfully implement the idea of interoperability, allowing electric vehicles to be charged using any inductive charger lacking any restrictions imposed by producers of the vehicles or chargers (Russer et al., 2013).

6.2.2. In-Motion IPT

With the help of in-motion charging technology, electric vehicles can now be charged while on the move, eliminating the inconvenience of having to make complete stops or endure long waiting periods. This technology provides numerous benefits, such as an extended driving range and a smaller on-board battery. Additionally, it supports the advancement of self-driving vehicles, enabling individuals to maximize their time behind the wheel. One method of achieving in-motion charging involves placing a power transmitter track below the road surface and connecting a power receiver coil with the vehicle chassis (Bentalhik et al., 2022). Both the transmitter and receiver coils are provided with high-frequency AC power. As the vehicle glides along its charging track, the electromagnetic fields produced by the transmitter seamlessly interact with the receiving coils to effectively transfer power. Nevertheless, the widespread adoption of in-motion charging is hindered by certain constraints. These include the considerable upfront expenses associated with installation, the necessity for exclusive charging lanes, which may not be practical in heavily populated urban areas, and the need for accurate alignment among the receiver and transmitter tracks to avoid power loss during transmission (Barsari et al., 2021) (Figure 6.3).

There are two types of transmitting tracks for in-motion charging: a stretched (single-long) track and a segmented track. Figure 6.4(a) illustrates the significant difference in length between the transmitter track and the coil upon the vehicle side in the context of a stretched configuration. The track length varies from about 1 m to 100 m underground, enabling the simultaneous charging of multiple electric vehicles on a single track (Onar, 2017). The extended transmitter is composed of a lengthy track constructed from litz wire, which provides power from a primary station.

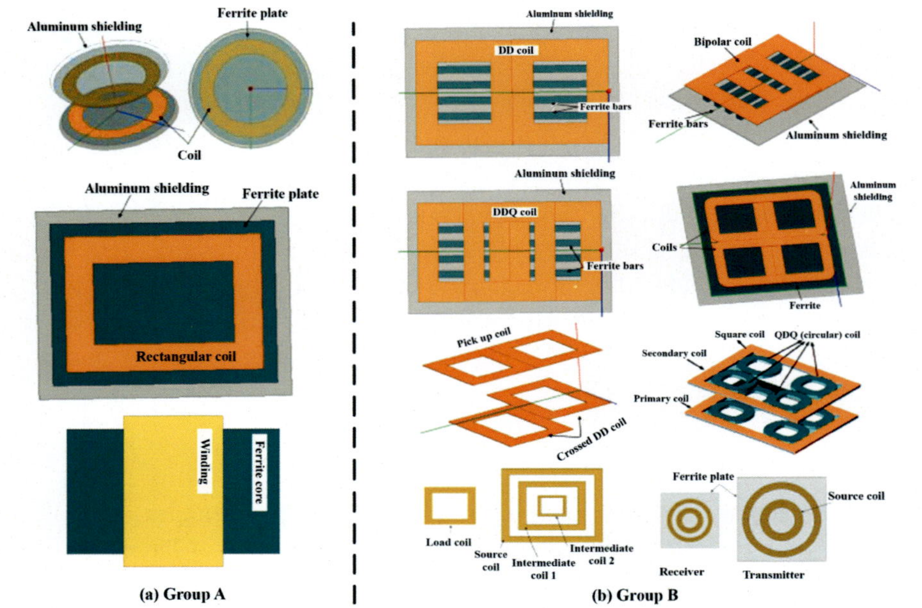

Figure 6.3. *Various designs for the transmitter and receiver stationary inductive charging pads: (a) Single-coil architectures; and (b) multiple-coil architectures.*

Source: Ahmed, A. S. Mohamed; Creative Commons License.

This station consists of a compensation network, an HF DC-AC converter (inverter), and an AC-DC converter (rectifier). A charging system utilizing a stretched track, referred to as an online electric vehicle (OLEV), was created specifically for general transport buses, and shuttle utilities in South Korea Nevertheless, this particular design has certain limitations. These include the potential for excessive electromagnetic field generation when misalignment occurs, higher system losses, the requirement for regular maintenance, and a lower coupling factor that results in reduced power and transmission efficiency. There are several classifications of the stretched transmitter track based on the type of magnetic core (Rahulkumar et al., 2022). These include the U-core type, E-core type, I-core type, ultra slim S-core type, S-core type, and cross-segmented type (X-track).

The charging track is composed of several coils that are placed in the ground and connected in either a parallel or series configuration, which can be seen in Figure 6.4(b) and (c). The segmented coils inside the track resemble the charging pad linked to the vehicle, creating a stationary charging system.

It also includes its compensation network, which is comparable to the resonate inductive power transfer system (RIPT).

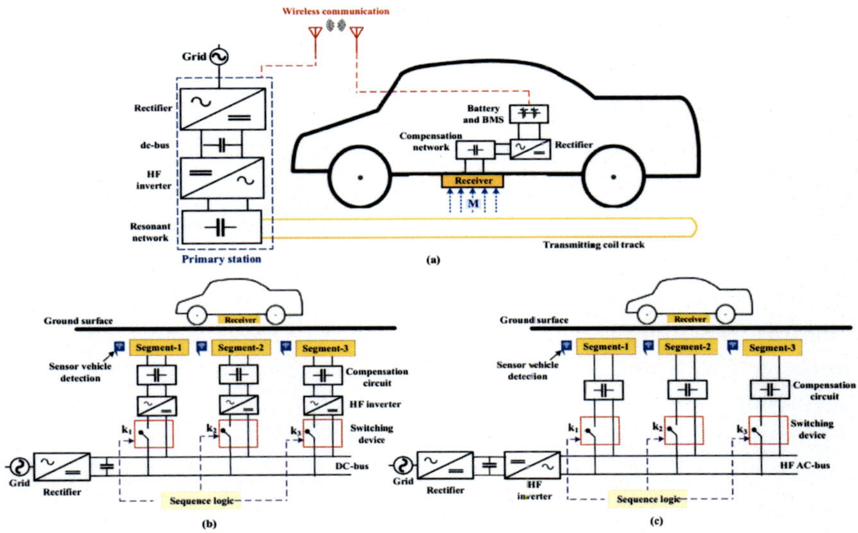

Figure 6.4. *Illustration of designs for in-motion inductive transmitter: (a) Single long coil track architecture; (b) segmented transmitter architecture using a common DC bus; and (c) segmented transmitter architecture using a common high-frequency AC bus.*

Source: Hamid Metwally. Mohamed; Creative Commons License.

Within this charging architecture, a single transmitter trackpad is activated as the electric vehicle passes over it (Apostoaia & Cernat, 2018). As the electric vehicle moves off the excited pad, the control system deactivates the excitation transitions to another pad and continues. This method enhances the efficiency of the system and minimizes any leakage EMFs. In addition, it enables the installation of separate compensation networks on each charging pad, which improves the durability of the entire track length. Nevertheless, the implementation of the segmented charging track necessitates the incorporation of numerous compensation circuits, converters, and coils for transmission, and conversion devices, resulting in a significant cost and complexity in its construction. It is important to address the issue of self-coupling among the transmitter segments by raising the horizontal distance among them (Shaier et al., 2023). This can result in power fluctuations and a noticeable decrease in power transmission if the receiver coincides with the middle of that distance. Minimizing the distance between the transmitter segments can help mitigate power fluctuations, although this may impact the resonant components as a result of coil self-coupling (TA & Marshiana, 2022).

6.3. COMPENSATION NETWORKS

Learning Objectives:

- Understand the role and benefits of compensation networks in inductive charging systems.

Compensation networks play a crucial role in inductive charging systems, providing a range of advantages. And address the issue of an important air gap distance, which leads to a high leakage inductance. This, in turn, enhances the power and efficiency of the system. In addition, lowers the apparent power demands by providing reactive power, achieves a power factor of unity (PF) alongside zero phase angle (ZPA) to avoid bifurcation analysis, and facilitates soft switching in electronic devices. Researchers from MIT conducted a study in 2007 that showed a significant increase in transmitting efficiency across an air gap distance when compensation elements were added to the charging circuit. This finding inspired the incorporation of these elements (Wang et al., 2021).

There are two main categories of compensation methods: basic (classical) compensation techniques and mixed (composite) compensation techniques. Compensation techniques commonly used in this field involve the connection of resonant capacitors to both the transmitter and receiver, as shown in Figure 6.5. There are four fundamental techniques available for consideration: parallel–series compensation (P–S), series–series compensation (S–S), series-parallel compensation (S–P), and parallel–parallel compensation (P–P). Various hybrid compensation methods are created by combining capacitors and inductors, as depicted in Figure 6.5. These techniques include CCL, LCL–P, LCL, LCL–LCL, S–CLC, LCCL–LCCL, LCC, and others (Zhu et al., 2016).

The majority of research in this field tends to concentrate on fundamental compensation circuits, while a few studies delve into the realm of hybrid compensation techniques. As an illustration, a three-phase bipolar pad demonstrated an impressive transmission rate of 95% while transmitting 50 kW through a 150 mm air gap with the aid of a series of compensation circuits. An analysis was conducted to enhance the transmission effectiveness of systems via asymmetric coils by exploring series-parallel and series-series compensation techniques. It was observed that the series-parallel compensation technique demonstrated higher efficiency with a larger transmitter coil compared to the receiver coil.

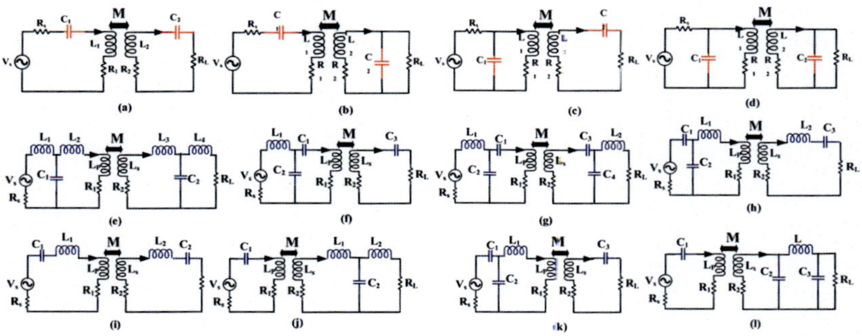

Figure 6.5. *Illustration of various compensation techniques: (a) Series–series; (b) series–parallel; (c) parallel–series; (d) parallel–parallel; (e) LCL–LCL; (f) LCC–S; (g) LCC–LCC; (h) CCL–LC; (i) LC–LC; (j) S–LCL; (k) CCL–S; and (l) S–CL.*

Source: Sameh I. Selem. Mohamed; Creative Commons License.

The choice of capacitance (C) and inductance (L) values is influenced by various factors, including the coupling factor, quality factor, and compensation technique on both sides. Parallel compensation is not commonly employed on the transmission side because of its highly intricate calculations, load reliability concerns, and input impedance (Deng et al., 2019). The capacitance used in compensation techniques relies on the transmitter's inductance (L1) and resonance frequency, whereas the mutual inductance among the transmitter and receiver coils also contributes to the process. The compensation techniques of P–P and P–S involve the use of capacitance on the transmitting side. The value of this capacitance is determined by factors such as inductance, coupling factor, and load impedance. These techniques necessitate a substantial operating voltage for the efficient transmission of power, owing to their significant input impedance. The techniques offer several advantages, such as high transfer efficiency, high power factor, and wide load variability, even at small mutual inductance. Which can operate at low power levels, enable long transmission distances, and utilize the same frequency for operation (Zhao et al., 2016).

Compensation circuits that combine different elements, including LCC–LCC and LCL–LCL, have been found to significantly enhance the efficiency of inductive wireless power charging systems, regardless of the level of coupling and loading. Nevertheless, the presence of leaking resistance in inductors and capacitors results in higher copper losses when compared to S-S compensation techniques. These both-sided methods (LCL and LCC) are well-suited for batteries for electric vehicle charging because they enable the present source of power at the vehicle side to work alongside an inverter of voltage on the transmitter. Through extensive research, the study focused on the technique of both-sided LCC compensation (Tian et al., 2016). It was discovered that by making adjustments to the LCC circuit, it is possible to achieve zero current switching and ensure a consistent root mean square (RMS) output current. One method to achieve the power factor of unity operation is by employing the LCC compensation method to compensate for the reactive power on the receiver side. Furthermore, this compensation technique remains unaffected by variations in loading conditions and coupling factors. An LLC compensation method was employed on the transmitter side, operating at frequencies above 1 MHz. This results in a significant increase in transmission effectiveness and a reduced VA rating (Fan et al., 2011).

6.4. CURRENT STATE OF THE ART OF IPT SYSTEM

Learning Objectives:

- Understand the current and state of the art inductive power transfer systems.

There are three different approaches to implementing inductive charging for electric vehicles. The first method is stationary inductive charging, involving charging the car while it is parked for an extended period at locations such as home, work, or parking lots. The second method is quasi-dynamic inductive charging, which allows the car to charge while it is moving at very low speeds or during brief stops, like at traffic lights or intersections. The third method is in-motion inductive charging, which enables the electric vehicle to charge while it is in motion at high speeds on highways and interstates (Kim et al., 2018).

6.4.1. Stationary Inductive Charging

The design of stationary charging systems for EVs is of utmost importance to meet precise power levels and air gap requirements. As a result, a significant amount of research and development is being dedicated to inductive charging systems. There are two categories for EV power levels, which are determined through global criteria and guidelines. There are two categories of wireless power charging. The first category is designed for light-duty electric vehicles (LDEVs) and has a power range of 1 to 22 kW. The second category can be used for medium- and heavy-duty electric vehicles (MDEVs/HDEVs) and can transmit power higher than 22 kW. The power requirements are also impacted by the dimensions and distance from the ground of each vehicle category (Mohamed et al., 2021). The air gap distance between the ground coil and vehicle coil within an inductive system can vary, with a range of 100 to 400 mm for LDEVs and even larger for MDEVs and HDEVs.

Practice Problem

An in-motion inductive charging system for electric vehicles faces challenges related to vehicle alignment during charging. If a vehicle misalignment reduces the charging efficiency to 80%, and the charging infrastructure intends to transfer 20 kWh of energy to the vehicle, how much energy is effectively received by the vehicle?

Solutions to Practice Problem

Given:
Charging efficiency due to misalignment = 80% = 0.80
Energy intended to be transferred by the charging infrastructure = 20 kWh
Effective energy received by the vehicle = Energy intended × Efficiency
Effective energy received = 20 kWh × 0.80
Effective energy received = 16 kWh
Therefore, the electric vehicle effectively receives 16 kWh of energy during charging despite misalignment issues.

Furthermore, vehicle manufacturers take an active role in the development of stationary inductively charged systems, either through independent efforts or in partnership with companies that specialize in electric car supply equipment. Their goal is to improve the characteristics of stationary inductive systems for charging and establish a solid basis for the global acceptance of this technology (Schoute, 2009).

6.4.2. In-Motion Inductive Charging

In 1976, an initial attempt was made to develop a dynamic inductive charging system capable of transmitting 8 kW of power. However, it did not achieve optimal efficiency. In 1979, the Santa Barbara electrical bus underwent development as another prototype. In 1992, the PATH project conducted experiments both in the lab and in real-world settings to test an in-motion charging mechanism. The design of the inductive charging system is developed and suspended in a bus (Barsari et al., 2021) (Table 6.1).

Table 6.1. Commercial Electric Vehicle Products for Stationary Inductive Charging System

Power (kW)	Supply Voltage (V AC)	Frequency (kHz)	Airgap (mm)	Efficiency
3.6	240	85	100–250	G2B: >90%
3.6–7.2	208–240	20	102	G2B: 90%
2–10	208–240	85	≤304.8	G2B ≥85%
3.6	240	85	160–220	G2B: >90%
50	277	23.4	165–190	DC to DC: 89%–92%
60–180	460	20	40	G2B: >90%
10	240	20	200	G2B: >92%
200	NA	20	203.2	G2B: 94%
25	240	20	241.3	G2B: 90%

Source: Sameh I. Selem. Mohamed; Creative Commons License.

Power transmission systems have been implemented on the roads, allowing for the examination and analysis of potential environmental impacts associated with the charging system. The design of the PATH project successfully transmitted an electrical power of 60 kW using a conversion efficiency of 60% over an air gap distance of about 76 mm. However, this initial version of the project faced significant challenges that hindered its commercial viability (Manshadi & Khodayar, 2020).

These included the bulky and heavy magnetic coupler, the expensive installation requirements, the need for high current because of an inadequate frequency of operation of up to 400 Hz, and the generation of high levels of acoustic noise. Furthermore, the power transmission via an air gap distance was not suitable for the ground clearance for electric vehicles. Despite the various challenges encountered in this project, it served as an initial catalyst for researchers and engineers to enhance and advance in-motion charging technology (Shaier et al., 2023).

6.5. CHALLENGES AND OPPORTUNITIES

Learning Objectives:

- Understand challenges in stationary inductive charging: efficiency, standardization, and infrastructure.
- Explore challenges in dynamic inductive charging: alignment, safety, and standardization.

Inductive charging presents a range of challenges including efficiency, positioning, heat management, compatibility, and cost. However, it offers an automatic and convenient solution for charging. It is of utmost importance to tackle the obstacles related to both stationary and dynamic inductive charging over electric vehicles. This will greatly enhance the charging experience, boost efficiency, and expedite the widespread use of electric vehicles within the transportation industry (Mohamed et al., 2020).

6.5.1. Challenges Associated with Stationary Inductive Charging for EV

6.5.1.1. Power Transfer Efficiency

Energy losses are commonly observed in inductive charging systems used in electric vehicles. These losses occur due to various factors, including the distance between the charging pad and the vehicle, alignment issues, and electromagnetic interference. Improving power transmission efficiency is crucial to reduce energy waste while achieving faster and more efficient charging (Lee & Chae, 2017).

6.5.1.2. High-Power Charging

Effectively achieving the transmission of high power wirelessly presents a notable obstacle in the realm of inductively charged batteries for electric vehicles. Addressing the constraints regarding technology efficiency and heat dissipation is crucial for achieving safe and efficient high-power charging, which in turn reduces charging time and enhances the usability of electric vehicles (Miller & Daga, 2015).

6.5.1.3. Standardization

Compatibility challenges within the inductively charging battery industry for electric vehicles arise due to the lack of standardized charging protocols. The lack of standardized charging protocols among automakers and charging infrastructure providers creates obstacles to seamless interoperability.

Encouraging the establishment of uniformity and coordination in charging systems is crucial to guarantee a seamless and convenient electrical charging procedure for electric automobile owners (Van Kampen & Zijlstra, 1961).

6.5.1.4. Infrastructure Deployment

Creating a widespread and readily available structure for stationary inductively charged devices is a difficult task. Implementing charging pads at various locations requires significant financial investments and coordination between multiple parties. Establishing a comprehensive charging network across a large geographic region is crucial for the widespread acceptance and convenient use of electric vehicles (Stephens et al., 2010).

6.5.2. Challenges Associated with Dynamic Inductive Charging for EV

Alignment and positioning refer to the arrangement and placement of elements in a specific order or location. Dynamic inductively charged systems in electric vehicles necessitate accurate alignment and positioning among the charging infrastructure integrated into the road and the receiver mounted on the vehicle. Ensuring the car stays properly aligned while in motion is difficult since any deviations can result in reduced charging efficiency and possible disconnection (Dimitriadou et al., 2023).

6.5.2.1. Power Transfer Efficiency

Dynamic inductive charging methods have additional efficiency obstacles when compared with stationary systems. As the vehicle traverses the infrastructure for charging, the arrangement and spacing of the charging pads may differ, which might impact the efficiency of power transfer. It is crucial to overcome these problems and provide a steady and efficient transmission of power for dynamic charging (Lee & Chae, 2017).

6.5.2.2. Safety and Reliability

It is crucial for dynamic inductive systems of charging to prioritize safety and reliability both for the infrastructure used for charging and the vehicle. It is of utmost importance to safeguard the charging pads integrated into the road from various environmental elements, including moisture, debris, and extreme temperatures. In addition, it is crucial to incorporate strong safety protocols to avoid any unintended electrical contact and maintain the system's reliability while in use (Duncan, 2000).

6.5.2.3. Standardization and Interoperability

Just like stationary inductive charging, dynamic charging systems encounter challenges when it comes to standardization and interoperability. The lack of uniformity and the use of different technologies by various manufacturers in dynamic charging systems create obstacles to achieving compatibility among charging infrastructure and electric vehicles. Facilitating widespread adoption and seamless integration of dynamic charging technologies can be achieved through the standardization of dynamic charging protocols and the promotion of interoperability (Lee et al., 2021).

6.5.2.4. Cost and Complexity

Addressing the expense of infrastructure poses a pressing challenge. Applying inductive stationary and dynamic charging technologies necessitates the setting up of charging pads along with backing infrastructure in a variety of locations, including offices, homes, public spaces, and roads. Installing inductive charging on a large scale is not without its challenges. The cost of setting up and maintaining the necessary infrastructure, along with the requirement for widespread use to make the investment worthwhile, present's significant obstacles (Welch & Waxman, 2003) (Figure 6.6).

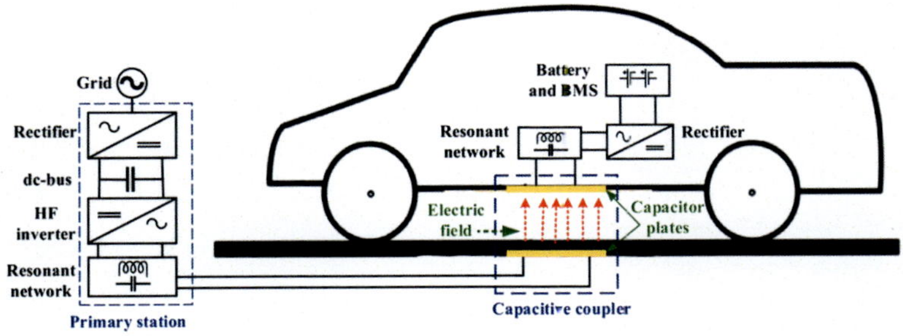

Figure 6.6. *Illustration of components of an exemplary capacitive power transfer system.*

Source: González, M.; Creative Commons License.

SUMMARY

- The chapter starts with a discussion on the design and functionality of inductive power transfer (IPT) systems for EV charging. It starts by dissecting magnetic coupler system architecture, detailing their components and operational principles in both stationary and in-motion scenarios.
- Compensation networks, crucial for optimizing IPT system efficiency, are discussed in depth. They regulate power transfer and ensure compatibility between charging infrastructure and EVs.
- The chapter describes the current state of IPT systems, showcasing advancements and emerging technologies shaping the landscape of wireless EV charging.
- Challenges and opportunities in both stationary and dynamic inductive charging are explored. Technical hurdles like power transfer efficiency, alignment optimization, and standardization are identified for stationary charging, while dynamic charging presents complexities in managing power transfer during vehicle motion and addressing safety concerns.

REVIEW QUESTIONS

1. Explain the magnetic coupler system architecture in wireless EV charging, detailing key components and functions.
2. Describe how compensation networks optimize IPT system efficiency and ensure compatibility between infrastructure and EVs.
3. Assess advancements and emerging technologies in IPT systems for EV charging.
4. Identify challenges in stationary inductive charging, including power transfer efficiency and standardization.
5. Discuss challenges and opportunities in dynamic inductive charging, focusing on power transfer management during vehicle motion.

MULTIPLE CHOICE QUESTIONS

1. What is the primary function of compensation networks in inductive power transfer (IPT) systems?

a. Managing vehicle alignment

b. Optimizing power transfer efficiency

c. Ensuring vehicle safety

d. Regulating charging speed

2. Which of the following is a key challenge associated with stationary inductive charging for electric vehicles?

a. Managing power transfer during vehicle motion

b. Aligning the vehicle with the charging infrastructure

c. Addressing safety concerns during charging

d. Optimizing charging speed

3. What aspect of IPT systems is emphasized in the discussion of magnetic coupler system architecture?

a. Vehicle compatibility

b. Power transfer efficiency

c. Structural components

d. Safety features

4. In the context of inductive wireless charging for EVs, what do compensation networks regulate?

a. Charging speed

b. Vehicle alignment

c. Power transfer

d. Safety protocols

5. What is a primary opportunity identified in the chapter regarding inductive wireless charging for electric vehicles?

a. Enhanced safety features

b. Increased charging speed

c. Widespread adoption of electric vehicles

d. Compatibility with all vehicle types

Answers to Multiple Questions

1. (b); 2. (b); 3. (c); 4. (c); 5. (c)

REFERENCES

1. Apostoaia, C., & Cernat, M. (2018). Inductive power charging for electric vehicles in misalignment conditions. *Annals of the University of Craiova, Electrical Engineering Series, 42*(1), 8–14.

2. Barsari, V. Z., Thrimawithana, D. J., & Covic, G. A. (2021). An inductive coupler array for in-motion wireless charging of electric vehicles. *IEEE Transactions on Power Electronics, 36*(9), 9854–9863.
3. Bentalhik, I., Lassioui, A., El Fadil, H., Bouanou, T., Rachid, A., El Idrissi, Z., & Hamed, A. M. (2022). Analysis, design and realization of a wireless power transfer charger for electric vehicles: Theoretical approach and experimental results. *World Electric Vehicle Journal, 13*(7), 121–499.
4. Buja, G., Rim, C. T., & Mi, C. C. (2016). Dynamic charging of electric vehicles by wireless power transfer. *IEEE Transactions on Industrial Electronics, 63*(10), 6530–6532.
5. Chakibanda, V., & Komanapalli, V. L. N. (2023). Optimization in magnetic coupler design for inductively coupled wireless charging of electric vehicles: A review. *Arabian Journal for Science and Engineering, 48*(11), 14257–14294.
6. Covic, G. A., & Boys, J. T. (2013). Inductive power transfer. *Proceedings of the IEEE, 101*(6), 1276–1289.
7. Deng, Z., Peng, X., & Qiao, Y. (2019). Residual compensation networks for heterogeneous face recognition. In *Proceedings of the AAAI Conference on Artificial Intelligence, 33*(1), 8239–8246.
8. Dimitriadou, K., Rigogiannis, N., Fountoukidis, S., Kotarela, F., Kyritsis, A., & Papanikolaou, N. (2023). Current trends in electric vehicle charging infrastructure; opportunities and challenges in wireless charging integration. *Energies, 16*(4), 2057.
9. Duncan, J. M. (2000). Factors of safety and reliability in geotechnical engineering. *Journal of Geotechnical and Geoenvironmental Engineering, 126*(4), 307–316.
10. Fan, L., Zhu, C., Miao, Z., & Hu, M. (2011). Modal analysis of a DFIG-based wind farm interfaced with a series compensated network. *IEEE Transactions on Energy Conversion, 26*(4), 1010–1020.
11. Hsieh, Y. C., Lin, Z. R., Chen, M. C., Hsieh, H. C., Liu, Y. C., & Chiu, H. J. (2016). High-efficiency wireless power transfer system for electric vehicle applications. *IEEE Transactions on Circuits and Systems II: Express Briefs, 64*(8), 942–946.
12. Hutchinson, L., Waterson, B., Anvari, B., & Naberezhnykh, D. (2019). Potential of wireless power transfer for dynamic charging of electric vehicles. *IET Intelligent Transport Systems, 13*(1), 3–12.
13. Kim, D., Abu-Siada, A., & Sutinjo, A. (2018). State-of-the-art literature review of WPT: Current limitations and solutions on IPT. *Electric Power Systems Research, 154*, 493–502.

14. Kim, S., Covic, G. A., & Boys, J. T. (2016). Tripolar pad for inductive power transfer systems for EV charging. *IEEE Transactions on Power Electronics, 32*(7), 5045–5057.

15. Lee, E., Seo, Y. D., Oh, S. R., & Kim, Y. G. (2021). A survey on standards for interoperability and security in the Internet of Things. *IEEE Communications Surveys & Tutorials, 23*(2), 1020–1047.

16. Lee, K., & Chae, S. H. (2017). Power transfer efficiency analysis of intermediate-resonator for wireless power transfer. *IEEE Transactions on Power Electronics, 33*(3), 2484–2493.

17. Lukic, S., & Pantic, Z. (2013). Cutting the cord: Static and dynamic inductive wireless charging of electric vehicles. *IEEE Electrification Magazine, 1*(1), 57–64.

18. Madawala, U. K., Neath, M., & Thrimawithana, D. J. (2011). A power–Frequency controller for bidirectional inductive power transfer systems. *IEEE Transactions on Industrial Electronics, 60*(1), 310–317.

19. Mahesh, A., Chokkalingam, B., & Mihet-Popa, L. (2021). Inductive wireless power transfer charging for electric vehicles–A review. *IEEE Access, 9*, 137667–137713.

20. Mahmud, M. H., Elmahmoud, W., Barzegaran, M. R., & Brake, N. (2017). Efficient wireless power charging of electric vehicle by modifying the magnetic characteristics of the transmitting medium. *IEEE Transactions on Magnetics, 53*(6), 1–5.

21. Manshadi, S. D., & Khodayar, M. E. (2020). Strategic behavior of in-motion wireless charging aggregators in the electricity and transportation networks. *IEEE Transactions on Vehicular Technology, 69*(12), 14780–14792.

22. Miller, J. M., & Daga, A. (2015). Elements of wireless power transfer essential to high power charging of heavy-duty vehicles. *IEEE Transactions on Transportation Electrification, 1*(1), 26–39.

23. Mohamed, A. A., Shaier, A. A., Metwally, H., & Selem, S. I. (2020). A comprehensive overview of inductive pad in electric vehicles stationary charging. *Applied Energy, 262*, 114584.

24. Mohamed, A. A., Wood, E., & Meintz, A. (2021). In-route inductive versus stationary conductive charging for shared automated electric vehicles: A university shuttle service. *Applied Energy, 282*, 116132.

25. Onar, O. C. (2017). Guest editorial special issue on wireless charging systems. *IEEE Transactions on Transportation Electrification, 3*(2), 301–302.

26. Rahulkumar, J., Narayanamoorthi, R., Vishnuram, P., Bajaj, M., Blazek, V., Prokop, L., & Misak, S. (2022). An empirical survey on wireless inductive power pad and resonant magnetic field coupling for in-motion EV charging system. *IEEE Access, 11*, 4660–4693.
27. Russer, J. A., Dionigi, M., Mongiardo, M., & Russer, P. (2013). A bidirectional moving field inductive power transfer system for electric vehicles. In *2013 11th International Conference on Telecommunications in Modern Satellite, Cable and Broadcasting Services (TELSIKS)* (Vol. 1, pp. 17–20).
28. Schoute, M. (2009). The relationship between cost system complexity, purposes of use, and cost system effectiveness. *The British Accounting Review, 41*(4), 208–226.
29. Shaier, A. A., Mohamed, A. A., Metwally, H., & Selem, S. I. (2023). New design of high-power in-motion inductive charger for low power pulsation. *Scientific Reports, 13*(1), 17838.
30. Stephens-Romero, S. D., Brown, T. M., Kang, J. E., Recker, W. W., & Samuelsen, G. S. (2010). Systematic planning to optimize investments in hydrogen infrastructure deployment. *International Journal of Hydrogen Energy, 35*(10), 4652–4667.
31. Stielau, O. H., & Covic, G. A. (2000). Design of loosely coupled inductive power transfer systems. In *PowerCon 2000. 2000 International Conference on Power System Technology. Proceedings (Cat. No. 00EX409)* (Vol. 1, pp. 85–90).
32. TA, A. R., & Marshiana, D. (2022). Charging of electric vehicle in-motion. *NCERC International Journal of Advanced Science, Engineering and Technology, 1*(1), 15–23.
33. Thai, V. X., Jang, G. C., Jeong, S. Y., Park, J. H., Kim, Y. S., & Rim, C. T. (2019). Symmetric sensing coil design for the blind-zone free metal object detection of a stationary wireless electric vehicle's charger. *IEEE Transactions on Power Electronics, 35*(4), 3466–3477.
34. Tian, Y., Nix, A. R., & Beach, M. (2016). On the performance of opportunistic NOMA in downlink CoMP networks. *IEEE Communications Letters, 20*(5), 998–1001.
35. Van Kampen, E. J., & Zijlstra, W. G. (1961). Standardization of hemoglobinometry II. The hemiglobincyanide method. *Clinica Chimica Acta, 6*(4), 538–544.
36. Wang, Y., Zhang, H., & Lu, F. (2021). Review, analysis, and design of four basic CPT topologies and the application of high-order compensation networks. *IEEE Transactions on Power Electronics, 37*(5), 6181–6193.

37. Welch, J. J., & Waxman, D. (2003). Modularity and the cost of complexity. *Evolution, 57*(8), 1723–1734.
38. Zhao, J., Cai, T., Duan, S., Feng, H., Chen, C., & Zhang, X. (2016). A general design method of primary compensation network for dynamic WPT system maintaining stable transmission power. *IEEE Transactions on Power Electronics, 31*(12), 8343–8358.
39. Zhu, Q., Wang, L., Guo, Y., Liao, C., & Li, F. (2016). Applying LCC compensation network to dynamic wireless EV charging system. *IEEE Transactions on Industrial Electronics, 63*(10), 6557–6567.

CHAPTER 7

WIRELESS POWER TRANSFER IN ON-LINE ELECTRIC VEHICLE

LEARNING OBJECTIVES

At the end of this chapter, readers will be able to:

1. Understand the historical development of wireless power transfer technology.
2. Know the market trends and commercialization challenges of wireless charging systems.
3. Explore the applications of wireless power transfer in automobiles.
4. Understand the mechanism of electromagnetic field-based power transmission.
5. Differentiate between inductive coupling and resonant magnetic coupling.
6. Explore the design and optimization of wireless power transfer systems for electric vehicles

INTRODUCTORY EXAMPLE

WiTricity Corporation is at the forefront of the WPT technology industry. Their systems employ magnetic resonance technology to facilitate efficient power transfer over various distances, making them ideal for both stationary and dynamic charging applications. In a bustling urban environment, the fleet that includes electrical buses often faces the challenge of frequent recharging, leading to potential delays and the need for larger batteries, which in turn can limit the passenger capacity of these buses. To tackle this problem, WiTricity has implemented its WPT system, which enables buses to power up while on the move or at specific stops without the need for physical connections. The system utilizes inductive coupling, employing primary coils that are integrated into the road infrastructure at bus stops and specific sections of the bus route. These coils are then connected to the city's power grid. The electric buses come with secondary coils installed underneath them. When a bus comes to a chosen charging point or passes over a section using embedded primary coils, a magnetic field is generated by the system. This magnetic field then causes an electric current to be produced in the secondary coils of the bus, resulting in the dynamic charging of its batteries without the need for physical connectors. This configuration enables buses to charge while operating their routes, minimizing periods of inactivity, reducing reliance on bulky batteries, and enhancing the overall efficiency and convenience of the public transportation system. WiTricity's wireless power transfer system revolutionizes electric bus charging, eliminating the need for regular stops and bulky batteries. This cutting-edge technology boosts operational efficiency and offers an easy option for public transportation.

UNIT INTRODUCTION

The broad concepts of WPT systems that utilize magnetic fields, specifically focusing on their application in online EVs (Ahn et al., 2011). Let's dive into the fascinating world of WPT and how it is revolutionizing the field of EVs. WPT revolutionizes the transmission of electrical energy by eliminating the need for physical connectors. This breakthrough technology opens up a world of possibilities for both dynamic and stationary charging solutions (Ko & Jang, 2013). The core principles of WPT technology are explored, which encompass inductive and resonant magnetic coupling techniques. In addition, this chapter explores the workings of WPT and the development of online EVs (Li & Mi, 2014).

7.1. OVERVIEW OF WPT TECHNOLOGY

Learning Objectives:

- Understand the historical background and evolution of Wireless Power Transfer (WPT) technology.
- Know the application of WPT technology in automobiles.

7.1.1. Wireless Power Transfer Technology

Ever since Guglielmo Marconi's groundbreaking invention of long-distance radio transmission, wireless telecommunications technology has become a ubiquitous part of everyday life. Wireless communication has become an essential aspect of our daily routines, and the majority of individuals today have no idea of a life without a cell phone or laptop. The advancement of wireless communication has been made possible, in part, by the progress in energy storage systems. Individuals desire freedom from the constraints of being tethered to a power source. Nevertheless, there is a desire for enhanced performance in electronic devices concurrently. The increasing demand for portable electronics with advanced computing capabilities has driven the advancement of lithium batteries with high energy density (Lu et al., 2017).

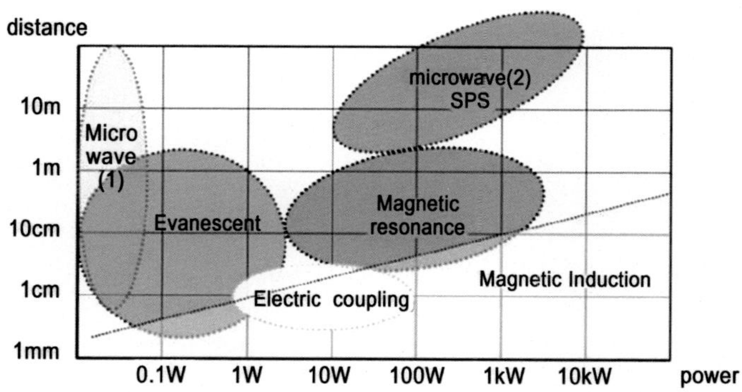

Figure 7.1. *Representation of the organizing wireless power transfer based on the amount of power transmitted and the distance covered.*

Source: Johnson, I.; Creative Commons License.

However, a new revolution in the field of wireless technology is emerging, aiming to eliminate the need for power transmission through physical wires. The concept of WPT traces back to the pioneering work of Nikola Tesla (1856–1943), who conducted experiments with the Tesla Tower in the early 1900s.

Tesla aimed to transmit power wirelessly through a waveguide connecting the earth's surface to the ionosphere (Shoki, 2013).

Despite the experiment's lack of success initially, researchers continued in their study of WPT technology. This ongoing research has led to the development of various applications, as illustrated in Figure 7.2. There were various applications including solar power satellites (SPS), cookers, magnetic induction for electric brush small power microwaves, and more (Triviño et al., 2021) (Figures 7.1 and 7.2).

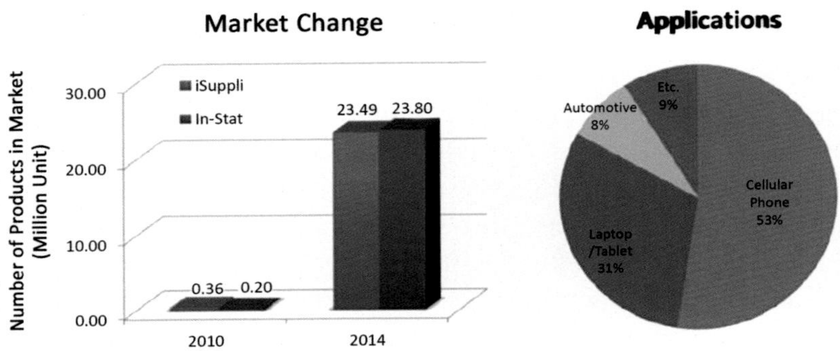

Figure 7.2. *Illustration of exploring the current market trend for wireless chargers.*

Source: González, M.; Creative Commons License.

7.1.2. Application to Automobiles

The development began with small portable electronic devices for consumers. However, while shifting focus to other electronic systems, a plethora of applications was discovered for utilizing WPT technology. Automobiles are one of the most significant applications. With the increasing integration of electronic devices in automotive vehicles, the application of WPT technology allows for the convenient transfer of power without the need for wires. Furthermore, the electronic gadgets utilized within the vehicle, such as smartphones or laptops, can be charged without the need for wires (Padmanabhan & Rao, 1993).

Transferring power to the vehicle is a highly complex technological endeavor. Typical electronic devices have a power output of less than 100 watts, making them fairly simple to put into effect and less challenging to get into the market. When considering the implementation of portable power transfer technology for vehicle power supply, one of the primary considerations is scaling up the power to beyond kilowatt levels. Addressing this issue goes beyond simply boosting the power supply. Various challenges arise in the realm of ferromagnetic materials, high-power semiconductor devices, power transfer efficiency, thermal considerations, magnetic field safety, weight and size, and cost. These concerns may have a notable impact on smaller power applications but have become even more crucial when dealing with high-power applications (Chandgude & Salunkhe, 2021). Despite the numerous challenges involved in implementing the WPT system for electric vehicles, it has garnered increased attention due to the growing demand for electric vehicles. In 1994, a research group from UC Berkeley, led by the Partners for Advanced Transit & Highways (PATH) project, obtained research findings on the technology of wireless power transfer (Oyama et al., 2008).

It has been demonstrated that a power of 10 Kilowatts per module could be efficiently transferred from the infrastructure built beneath the road, with a power efficiency of 60%. When the supply current was set at 2,000 amperes, the distance between the road surface and the bottom of the vehicle, known as the air gap, measured 7.5 cm. The vehicle is engineered to receive power wirelessly, whether it is stationary or in motion. The approach taken in this research was logical and served as a driving force for subsequent studies. However, the limited air gap proved insufficient to ensure a comfortable driving experience on actual roads. For instance, all vehicles have an air gap of at least 12 cm, with many vehicles having a larger air gap due to speed bumps. In addition, the current of 2,000 amperes is significant in generating a powerful magnetic field for wireless power transfer. However, this also posed a challenge as it exceeded the magnetic field regulations implemented in several countries. Cost reduction and audible noise problems were also left to solve (Young & Finlayson, 1976) (Figure 7.3).

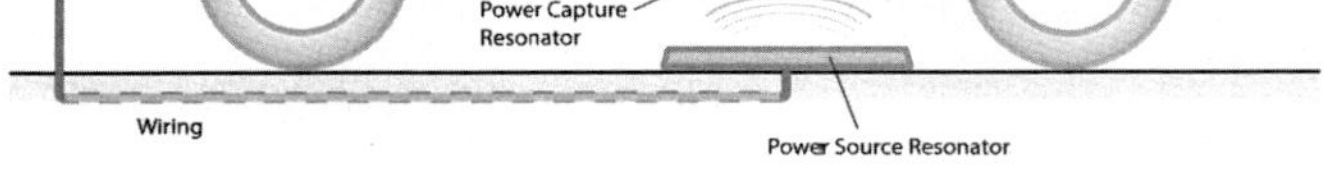

Figure 7.3. *Illustration of developing a cutting-edge wireless charging method for automotive that eliminates the need for plugs or charging cords.*

Source: Hidetochi Matsuki; Creative Commons License.

7.2. MECHANISM OF WIRELESS POWER TRANSFER

Learning Objectives:

- Understand the principles of electromagnetic fields and their role in wireless power transmission.
- Explore the concepts of inductive coupling and resonant magnetic coupling for efficient power transfer.

7.2.1. Magnetic Field and Electric Field

There are various methods by which power can be transferred. In the realm of electric power, the flow of electrons via conductors or semiconductors facilitates the transfer of energy between connected transmitters and receivers. When the transmitter and receiver have electrically divided, power can be transferred wirelessly using the electromagnetic field as a medium of transmission. Based on Maxwell's equations, it is observed that an electric field that changes over time produces a magnetic field, while a magnetic field that changes over time generates an electric field. The relationship between the magnetic field and the electric field is described as wave impedance. When measuring the magnitude of the electric field and magnetic field at a distant observing point from the source, one finds that the ratio of the magnetic field to the electric field will eventually reach a constant value known as wave impedance (Lottermoser et al., 2004).

$$\begin{aligned} \nabla \times E &= -\frac{\partial B}{\partial t} \\ \nabla \times H &= J + \frac{\partial D}{\partial t} \\ \nabla \cdot D &= \rho \\ \nabla \cdot B &= 0 \end{aligned} \qquad (7.1)$$

The properties of a field are influenced by various factors, including the origin of the field, the medium in which it exists, and the distance that exists within the source and the point of observation. Close to a source, the field characteristics are mainly influenced by the source traits. When the field travels a significant distance from its source, its characteristics are primarily influenced by the medium it is passing through. Thus, the area surrounding a radiation source can be divided into two separate regions. When the point of observation is near its source, the field falls within the near-field region. When the point of observation is located at a significant distance from its source, the field is said to be in the far-field region (Koch, 2000). If certain conditions are met, the region can be classified as a transition region.

$$\text{Distance} = \frac{\lambda}{2\pi}. \qquad (7.2)$$

The value for lambda is directly related to the wavelength, so the nearby region refers to an area in which the distance is significantly smaller compared to the wavelength (Weisheit et al., 2007).

In Figure 7.4, when the standardized distance is significantly greater than 1, it indicates that the system lies within the far-field region. At this point, the wave impedance converges to approximately 377 ohms. If the standardized distance is significantly less than 1, it falls within the near-field region. This region can be divided into two conditions: one where the electric field is dominant and another where the magnetic field is dominant. When a dipole antenna is utilized to generate an electromagnetic field, the electric field takes precedence, and the wave impedance exceeds 377 ohms. When utilizing a loop antenna, the magnetic field takes precedence and results in a wave impedance that is below 377 ohms. Typically, portable charging for electric vehicles relies on the utilization of the near field region under magnetic field predominant conditions (Matsukura et al., 2015).

Power transmission can utilize both the electric field and the magnetic field. However, the majority of applications use magnetic field transmission. This is because magnetic fields have the advantage of experiencing less attenuation and reflection, due to the high dielectric material. Additionally, magnetic fields are less influenced by permittivity. Furthermore, the implementation of the magnetic coupling structure can be more straightforward compared to electric field coupling. In Figure 7.5 it is shown that the loop antenna is capable of generating and receiving the magnetic field.

Additionally, the transmission and reception of high power can be achieved through the use of large wires and the magnetic field. For effective electric field coupling, it is necessary to have a significant capacitance between the transmitter and receiver. This means that instead of using dipole antennas with wires, larger plates are needed in both the transmitter and receiver (Conrad, 2000).

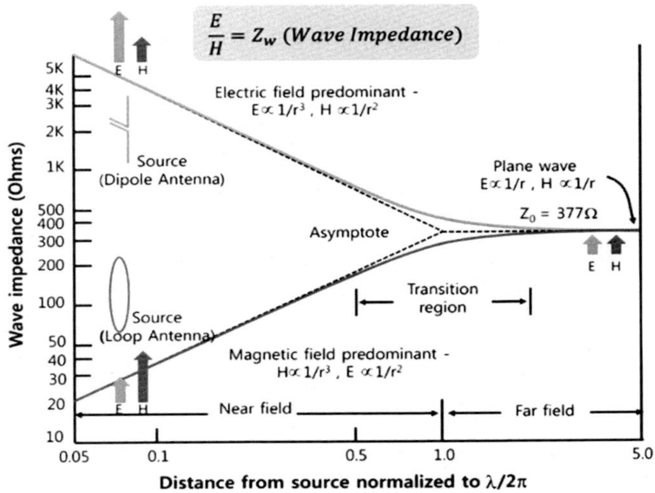

Figure 7.4. *Illustration of wave impedance as a function of the distance from the source.*

Source: Kurs, A.; Creative Commons License.

7.2.2. Exploring Inductive Coupling and Resonant Magnetic Coupling

Inductive wireless power transmission has been utilized for an extensive period, primarily in the context of electric toothbrushes. Inductive coupling technology is utilized in electric toothbrushes to recharge their batteries, which are frequently exposed to water. This particular wireless charging system falls under the category of inductive coupling. Inductive coupling typically involves the transmission of wireless power over a short distance, usually just a few millimeters, from the transmitter coil to the receiver coil (Hoult & Tomanek, 2002) (Figure 7.5).

Due to the proximity of the coils, leakage of the magnetic field is minimal, resulting in a remarkably high efficiency ranging from 60% to 90%. One drawback of inductive coupling is the challenge of increasing the distance, as well as the need for precise alignment between the transmitter and receiver. Failing to meet these conditions results in a decline in the transfer power (Shinohara, 2012).

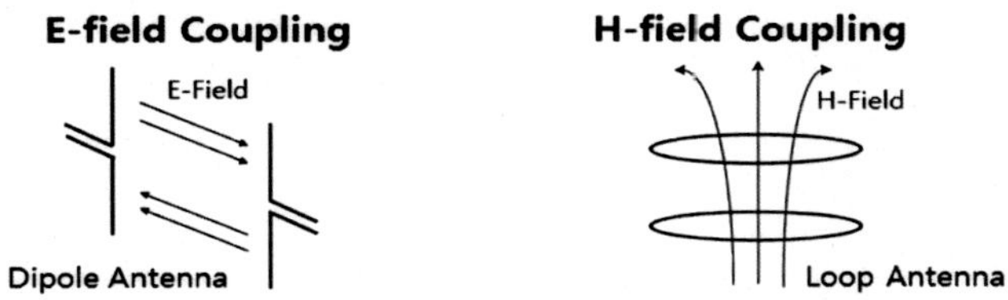

Figure 7.5. *Illustration of exploring the transmission and reception of electric and magnetic fields through the use of dipole and loop antenna.*

Source: Moffatt, R.; Creative Commons License.

Practice Problem

A 1500 kg vehicle traveling at 80 km/h needs continuous wireless charging via inductive coupling. With an efficiency of 90% and a charging station outputting 20 kW, what is the minimum distance required for uninterrupted charging.

Solutions to Practice Problem

The force exerted by the inductive system must counteract friction and air resistance to maintain speed. Calculating this force, and considering the system's efficiency, we find the effective power transferred. Equating this with the force and solving for distance, we determine a minimum distance of approximately 548.75 meters.

The challenges of distance and alignment accuracy might be overcome by implementing a resonance circuit of the transmitter and receiver systems, known as resonant magnetic coupling. This concept is thoroughly explained using the coupled-mode theory. Furthermore, various studies have demonstrated the validity of this concept across a wide range of frequencies and power levels (Kuipers et al., 2012).

7.3. DESIGN OF ONLINE ELECTRIC VEHICLE

Learning Objectives:

- Understand the necessity and development of On-Line Electric Vehicles (OLEVs).
- Understand the design principles and optimization criteria for wireless power transfer systems.
- Explore electromagnetic field reduction technologies in OLEVs.

7.3.1. The Importance of On-Line Electric Vehicles

The development of the OLEV On-Line Electric Vehicle began in 2009 to tackle two pressing global issues: resource depletion and environmental pollution. Extensive research has been conducted on fully electric cars for a considerable period. However, there are still significant challenges that need to be addressed for a battery-powered electrical transportation system. It is widely recognized that electricity offers a more cost-effective and environmentally friendly alternative to petroleum resources. Several prominent automotive companies have made significant strides in the development of electric vehicles. However, there are still certain challenges that need to be overcome, such as the concerns surrounding increased size and weight, as well as the high cost of batteries. In addition, the depletion of lithium reserves may lead to a surge in prices, potentially resulting in the exclusion of electric vehicles from the automotive industry. Although there may be an increase in battery capacity, it is important to address other challenges such as lengthy recharging times, safety considerations, limited access to charging stations, and the associated construction expenses (Jang et al., 2015) (Figure 7.6).

Figure 7.6. *Concept of electric vehicles that operate online.*

Source: Fisher, P.; Creative Commons License.

KAIST Korea Advanced Institute of Science and Technology has developed the OLEV, an innovative electric vehicle that addresses battery issues by utilizing wireless charging technology from road-embedded infrastructure. The pick-up coils (receiver coils) within OLEV continuously receive and recharge their power via power lines (transmitter coils) incorporated beneath the road.

Due to its lower battery capacity, this type of electric vehicle can help reduce the price and weight of the vehicle, in addition to the expense of the building of power stations (Ko & Jang, 2013). The idea of WPT for EVs, in which energy is transmitted effortlessly while the vehicle is stationary or in motion, was first introduced two decades ago. Nevertheless, the development of the wirelessly operated electric vehicle as an alternative to traditional fuel-engine automobiles has presented significant challenges due to the rigorous demands involved (Park et al., 2015).

7.3.2. Topology Analysis

To simplify the theory and provide a clear explanation, it is common to utilize both LC resonant circuits using mutual inductance, as represented in Figure 7.7. The transmitter coil and receiver coil possess mutual inductance (M) to wireless power transmission, alongside their self-inductances. Incorporating two capacitances (C1 and C2) into the transmitter and receiver circuits effectively nullifies the self-inductance for the inductance (Mahadevan et al., 2006).

As a result, the overall impedance of each loop is reduced. To achieve the highest possible impedance reduction, it is crucial to ensure that the frequency of resonance in the LC resonance matches the frequency for the source (I1) (Magoni & Pansiot, 2001). Consequently, the imaginary component of a loop minimizes and the:

1. Induced Voltage:

$$V_2 = j\omega M i_1 \tag{7.3}$$

2. Efficiency:

$$K = \frac{\omega^2 M^2}{R_1 R_2 + \omega^2 M^2} = \frac{1}{1 + \frac{R_1 R_2}{\omega^2 M^2}} = \frac{1}{1 + \frac{1}{Q_{M1} Q_{M2}}} \tag{7.4}$$

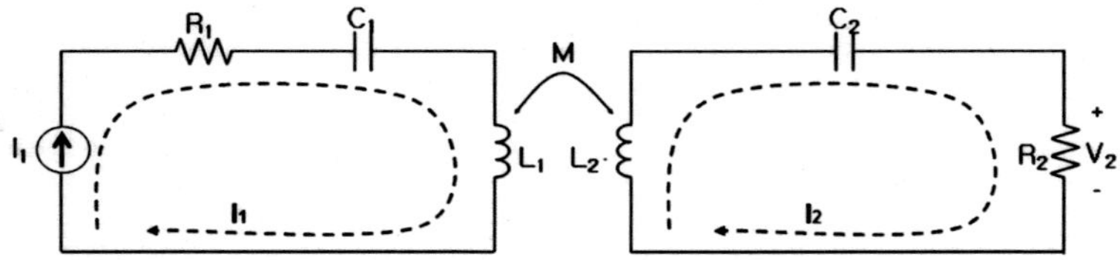

Figure 7.7. *Illustration of a simplified equivalent circuit is used to represent the current-source wireless power transfer system.*

Source: Tina Teng; Creative Commons License.

7.3.3. Coil Design for Electric Vehicle

Figures 7.8 and 7.9 depict a vertical magnetic flux form of power lines along with pickup coils. Underneath the road surface, two power lines have opposite current directions, creating a current loop. As a result of the electrical current flowing through the power lines, a magnetic flux is generated around each of the power lines (Mou et al., 2019). The magnetic fluxes generated by the two electric lines combine at the location between them, resulting in the cancellation of the magnetic fluxes outside of the electric lines.

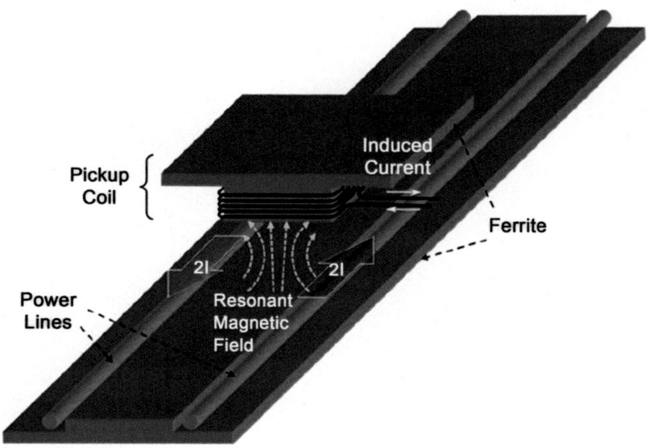

Figure 7.8. *Illustration of a perspective view of power lines and a pickup coil using vertical magnetic flux.*

Source: Henry, W.; Creative Commons License.

The pickup coil captures a vertical flux of magnetic energy by encircling the ferrite core with copper coil loops. The ferrite core plays a crucial role in implementing the shaped magnetic fields in resonance (SMFIR) technology, which is essential for designing wireless power transfer in On-Line Electric Vehicle (OLEV) systems. Its primary function is to optimize power transfer while minimizing electromagnetic field (EMF) emissions. One benefit of using this particular power line and pickup coil combination is that it allows for efficient power transfer. This is due to the alignment of the magnetic flux coming from the power lines with the flux to the pickup coil (Mohamed & Mohammed, 2018).

Figures 7.10 and 7.11 illustrate the horizontal flux kind of power lines and pickup coils. In this particular type, four power lines create two current loops. In Figure 7.9, there are 3 magnetic flux loops due to the close placement of two center conductors. The copper pickup coils wrapped around a horizontal ferrite core capture the horizontal magnetic flux, producing electricity to power vehicles (Zhang et al., 2016).

One drawback of this type has to do with the transfer power being lower compared to the vertical flux type due to the perpendicular directions of the induced magnetic field in the power lines or pickup coils of wire. One drawback is the challenge of increasing the size of a pickup coil loop, as doing so would also increase the overall thickness of the pickup coil. Nevertheless, this particular configuration of the power line and pickup coil results in a magnetic field that is relatively weaker on the side of the vehicle (Ahmad et al., 2019).

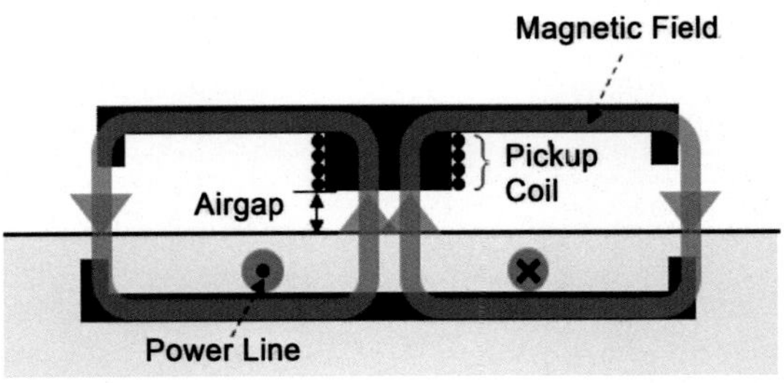

Figure 7.9. *Illustration of a cross-sectional view of power lines and a pickup coil with vertical magnetic flux.*

Source: Sunkyu Kong; Creative Commons License.

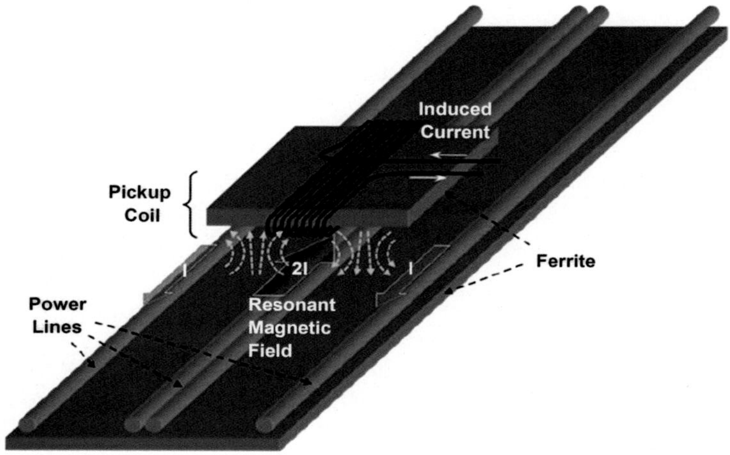

Figure 7.10. *Illustration of power lines with horizontal magnetic flux and a pickup coil shown in a perspective shot.*

Source: Henry, W.; Creative Commons License.

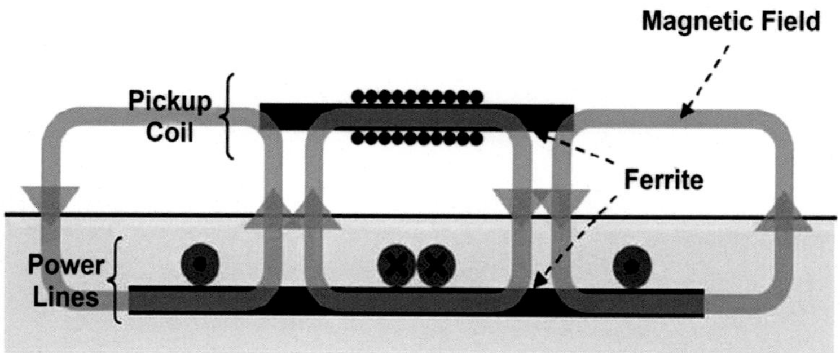

Figure 7.11. *Illustration of illustration depicting the cross-sectional view of power lines with horizontal magnetic flux and a pickup coil.*

Source: Bumhee Bae; Creative Commons License.

7.3.4. Electromagnetic Field Reduction Technology

Here is a visual representation of the magnetic flux density distribution in OLEV, as shown in Figure 7.12. For the vertical magnetic flux type, a single magnetic flux path exists between the power lines and pickup coils to facilitate the transfer of power. The return flux is directed towards the power lines through the sides in its primary path (Liu et al., 2012). In the horizontal magnetic flux type, there are two magnetic flux paths. These types of power lines include return flux paths located alongside the main flux path. These particular power lines include return flux paths located alongside the main flux path. The bordering magnetic flux is generated by the return flux path, and the level of this flux is determined as the EMF of OLEV (Wang et al., 2018). According to the regulations set by the Korea Communications Commission, the target electromagnetic field (EMF) level for On-Line Electric Vehicles (OLEV) in this study is 6.25 μT.

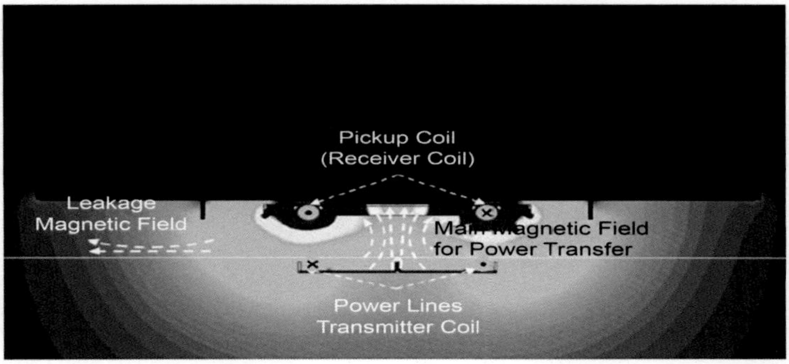

Figure 7.12. *Illustration of magnetic field distribution for OLEV.*

Source: Hiroshi Hirayama; Creative Commons License.

The power supply system of OLEV generates a significant magnetic field to transfer the required 100 kW of power to the vehicle. This results in thousands of μT of magnetic flux among the power beneath the vehicle lines and pickup coils during power transfer (Johansson, 2009). Therefore, if a mere 0.1% of the leakage magnetic field escapes from the OLEV system, it has the potential to surpass the 6.25 μT regulation for EMF levels.

Metal plates are used to passively shield the initial test version of OLEV to minimize the electromagnetic field (Yu et al., 2024). To ensure the safety of passengers, a sheet of metal is installed on the underside of the vehicle to shield against magnetic fields. Vertically plated shields are utilized to mitigate the magnetic field generated by power lines, as depicted in Figure 7.13. To enhance the shielding efficiency of a passive shield, metal brushes are used to create soft contacts between the bottom plate and the vertical ground plate, as illustrated in Figure 7.14. The metal brush consists of a collection of slender metal wires that are fastened below the bottom plate. It serves to establish a connection between the current within the vehicle body and the ground plate below the road surface. Figure 7.15 displays the photograph of the metal brush that has been implemented (Liu et al., 2012). Increasing the total amount of connections utilizing metal brushes can greatly enhance the shielding efficiency for passive shielding. The level of EMF has been reduced by about 14.4 μT towards 3.5 μT as the number of connections via metal brushes has risen from 2 to 8.

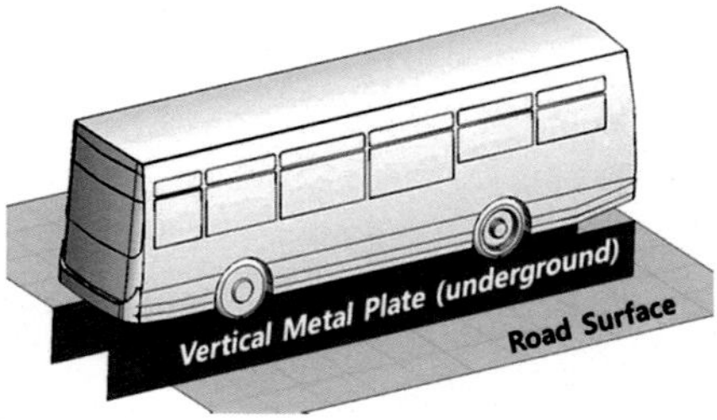

Figure 7.13. *Illustration of utilizing a vertical ground plate that is buried underground to minimize the impact of magnetic fields.*

Source: Hongseok Kim; Creative Commons License.

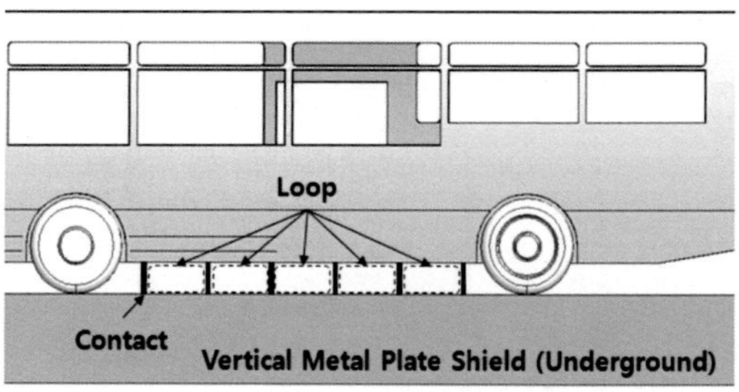

Figure 7.14. *Illustration of exploring the links between the vehicle body and an underground vertical plate of metal for passive shielding.*

Source: Hongseok Kim; Creative Commons License.

Nevertheless, the application of this technology is restricted to certain specific scenarios, as the metal brush is prone to wear and tear from friction. Additionally, the metallic ground shield on the road surface may become obstructed by non-metal materials, thereby disrupting the connection between the vehicle body surface and the ground plate (Markov, 2007).

Utilizing active shielding, either with or with no passive shields, can effectively reduce the electromagnetic field (EMF). Figure 7.16 illustrates the fundamental concept of an active shield. Just like power lines, the active shield is also a metal wire that carries the same frequency as the current, but in the opposite phase compared to the current in the pickup coil (Kheifets et al., 2005).

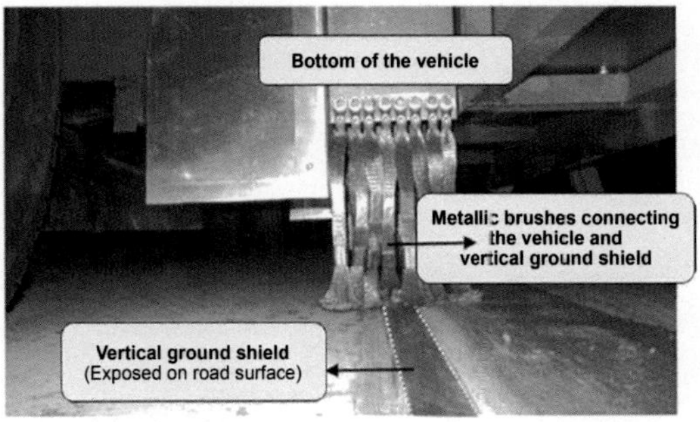

Figure 7.15. *Illustration of the metal brush has been incorporated into the passive shield found on the bottom of the OLEV.*

Source: Feliziani, M.; Creative Commons License.

When designing an active shield, it is crucial to carefully consider the path of magnetic fields from both the source and the active shield. Figure 7.17 shows the direction of the magnetic field. To ensure that the EMF level remains within the regulatory limits at all positions, it is crucial for the magnetic field emitted by the active shield to closely match that produced by the pickup coils at every position (Xanthakis et al., 2013).

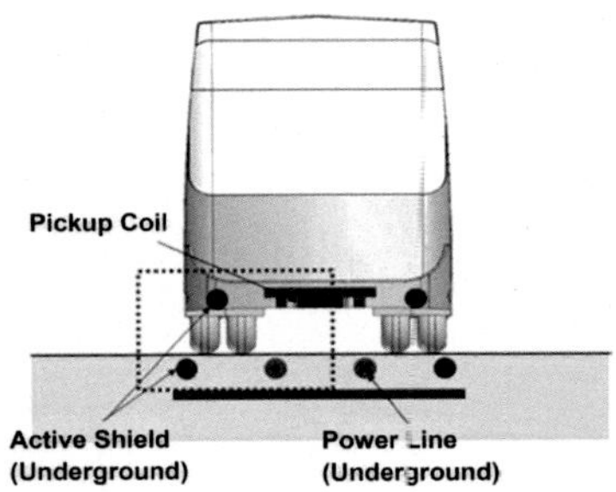

Figure 7.16. *Illustration of the exploring the idea of an active shield over OLEV.*

Source: Tina Teng; Creative Commons License.

When the distance from the road surface exceeds 20 cm, the magnetic field's vector aligns itself with the metal plate due to the presence of a metallic shield underneath the vehicle. Thus, positioning the active shield nearby for the pickup coil yields greater effectiveness. However, whenever the active shield is positioned closer to the pickup coil, it is expected that the current flow within the active shield will need to be increased (Schoenbach et al., 1997). Due to this factor, the positioning of an active shield becomes compromised when considering the effectiveness of the shielding and the magnitude of the current. In Figure 7.18, the density of magnetic flux is significantly reduced to just 1/10 of its original density when the active shield is implemented at the optimal current value.

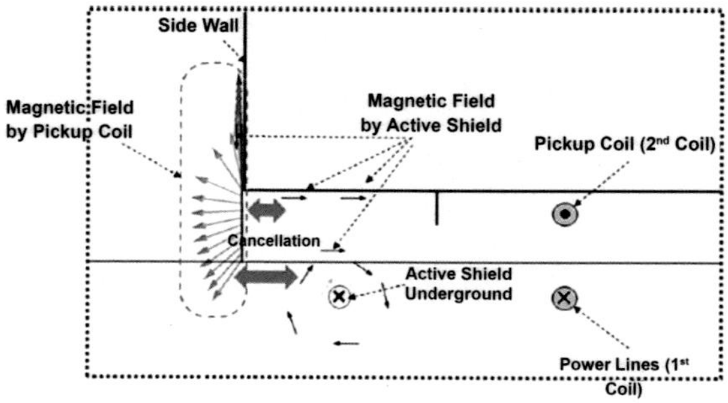

Figure 7.17. *Illustration of the understanding of the magnetic field from pickup coils and active shield.*

Source: Fuina, V.; Creative Commons License.

Figure 7.18 displays the EMF measurements following the implementation of an active shield on the OLEV. For the measurement, the chosen standard method is the IEC 62110, which determines the magnetic field measurements of electric systems. As per IEC 62110, the final measurement is determined by taking the mean value of measurements at three points (Tao & Tang, 2014). The use of active shield technology has effectively reduced the leakage of the magnetic field from 10.3 μT to 5.3 μT (Figures 7.19 and 7.20).

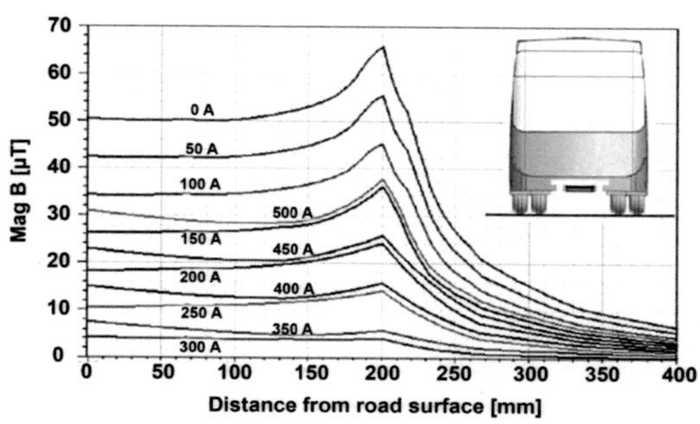

Figure 7.18. *Representation of the simulated variation of the electromotive force (EMF) level is analyzed to the electrical current of the active shield.*

Source: Yangbae Chun; Creative Commons License.

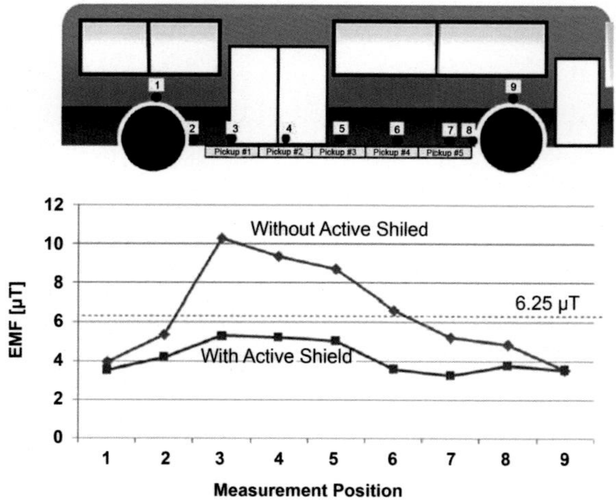

Figure 7.19. *Illustration of EMF levels in OLEV: Measured levels with and lacking active shield.*

Source: Youngjin Park; Creative Commons License.

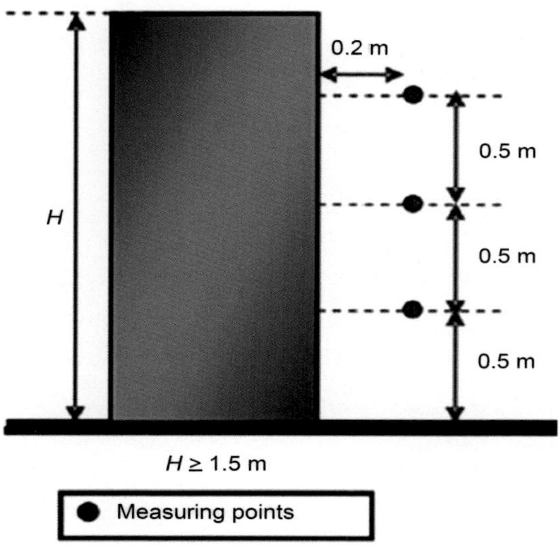

Figure 7.20. *The three positions of EMF measurement in IEC 62110.*

Source: Arthur Lu; Creative Commons License.

7.3.5. Design Procedure and Optimization

When designing the power lines and pickup coils structure for the OLEV system, there are three key factors to consider for the electrical efficiency of the wireless power transfer systems: power transfer capability, power transfer efficiency, or electromagnetic field leakage (Sundaresan et al., 1995).

The transfer of power capability defines the highest speed and battery recharging time for a vehicle. It refers to the maximum power that can be sent to the power lines beneath the road for the load inside the vehicle (Zakian & Kaveh, 2024). Based on a simplified equivalent circuit model for a wireless power transfer system in two series resonant coils, the power at the load RL can be determined. This calculation takes into account the frequency, and mutual inductance, along with the magnitude of the source current. Assumptions are made that the entire system operates at the resonance frequency, as indicated by Eqn. (7.5) (Lee et al., 2013).

$$P_C \cong \frac{\omega^2 M^2}{(R_2+R_L)^2+\left(\omega L_2-\frac{1}{\omega C_2}R_L\right)} I_1^2 R_L \cong \frac{\omega^2 M^2}{R_L} I_1^2 \tag{7.5}$$

$$P_C \cong \frac{\omega^2 M^2 R_L}{(R_2+R_L)^2+\omega^2 M^2 (R_2+R_L)} I_1^2 R_L \cong \frac{1}{1+\frac{R_1 R_L}{\omega^2 M^2}} \tag{7.6}$$

Having a high-power transfer rate is crucial for the successful commercialization of electric vehicles, especially when compared to other kinds of vehicles. To enhance efficiency, it is crucial to reduce losses at every stage of a power system for OLEV. Thanks to the advancements in power components working at 20 kHz, which were not available decades ago, the efficiency of an inverter has seen a remarkable increase. Furthermore, it is crucial to enhance the mutual inductance while minimizing the parasitical resistance of R1 and R2. By following the derivation in Eqn. (7.6), the efficiency can be further improved (Beiranvand et al., 2012).

The third criterion for leakage electromagnetic fields is directly related to the current's strength and inversely related to the distance from the current's location and the measurement position, assuming no shield is present. Nevertheless, when passive and active shields are applied, the magnitude of EMF undergoes significant changes, necessitating a separate design for the EMF (Ashayeri & Gelders, 1985).

Figure 7.21 illustrates the design procedure used by the WPT system in OLEV. During the initial design phase, it is crucial to establish the topology and dimensions of the physical structures. This includes determining the number of coils, the size and dimensions of the coils, and the placement of the ferrite core (Sobieszczanski et al., 1997). Once these physical dimensions are set, the mutual inductance is roughly determined, making it difficult to significantly alter its value in later stages. As a result, the design options are restricted, and achieving the optimal final design parameter becomes challenging (Phadke & Dehnad, 1988). When designing a WPT for an OLEV, it is crucial to take into account multiple parameters simultaneously during the design stage. Thus, the optimization stage becomes crucial once the design parameters and performance parameters have been established, as depicted in Figure 7.22.

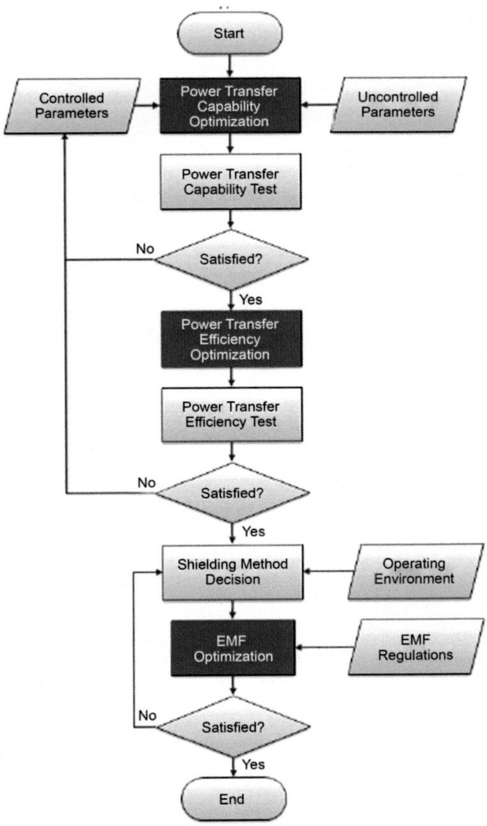

Figure 7.21. *Illustration of previous procedure of OLEV design.*

Source: Kyoungchoul Koo; Creative Commons License.

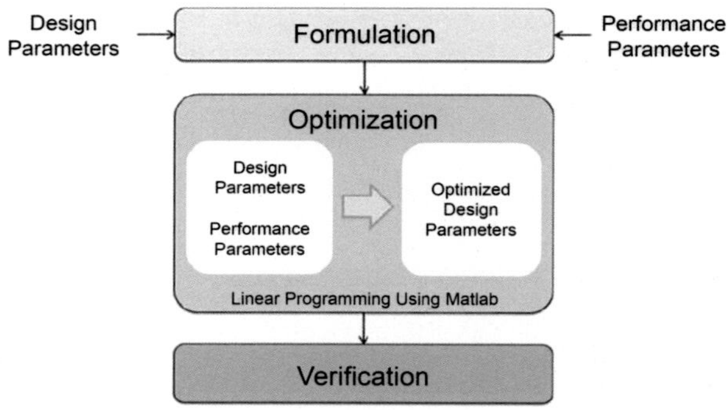

Figure 7.22. *Optimized procedure for OLEV design.*

Source: Junggun Byun; Creative Commons License.

SUMMARY

- The chapter explores WPT technology for electric vehicles (EVs), focusing on inductive and resonant magnetic coupling methods. These technologies enable dynamic and stationary charging without physical connectors, enhancing EV convenience and efficiency.
- The chapter discusses the integration of WPT systems in vehicles and infrastructure, emphasizing coil design, topology analysis, and electromagnetic field management to ensure efficient and safe power transfer.
- It highlights the necessity of On-Line Electric Vehicles (OLEVs) to address EV limitations like range and charging time, offering continuous charging solutions that reduce battery size and improve operational efficiency.

REVIEW QUESTIONS

1. Explain the basic principle of Wireless Power Transfer (WPT) technology.
2. Describe the differences between inductive coupling and resonant magnetic coupling in WPT systems.
3. What are the main advantages of using WPT technology in electric vehicles?
4. How does the magnetic field play a crucial role in WPT systems compared to the electric field?
5. Why is continuous charging important for the efficiency and practicality of electric vehicles?
6. What are the key factors to consider in the coil design for WPT systems in electric vehicles?
7. What are some potential environmental and economic impacts of widespread adoption of OLEVs?

MULTIPLE CHOICE QUESTIONS

1. What is the primary benefit of Wireless Power Transfer (WPT) technology for electric vehicles?

a. Lower manufacturing costs

b. Enhanced battery capacity

c. On-the-go charging

d. Reduced vehicle weight

2. Which technology is commonly used to minimize electromagnetic interference in WPT systems?

a. Capacitive filtering

b. Shielding

c. Antenna alignment

d. Optical isolators

3. Which component is essential for generating the electromagnetic field in a WPT system?

a. Battery

b. Primary coil

c. Transformer

d. Resistor

4. The necessity of On-Line Electric Vehicles (OLEVs) is driven by the need to address which limitation of conventional EVs?

a. High manufacturing cost

b. Limited driving range and long charging times

c. Excessive noise levels

d. Heavy environmental impact

5. Topology analysis in WPT systems primarily involves which aspect?

a. Color coordination

b. Arrangement and connection of WPT components

c. Developing user interfaces

d. Marketing strategies

Answers to Multiple Questions

1. (c); 2. (b); 3. (b); 4. (b); 5. (b)

REFERENCES

1. Ahmad, A., Alam, M. S., & Mohamed, A. A. (2019). Design and interoperability analysis of quadruple pad structure for electric vehicle wireless charging application. *IEEE Transactions on Transportation Electrification, 5*(4), 934–945.

2. Ahn, S. Y., Chun, Y. B., Cho, D. H., & Kim, J. H. (2011). Wireless power transfer technology in on-line electric vehicle. *Journal of the Korean Institute of Electromagnetic and Science, 11*(3), 174–182.
3. Ashayeri, J., & Gelders, L. F. (1985). Warehouse design optimization. *European Journal of Operational Research, 21*(3), 285–294.
4. Beiranvand, R., Rashidian, B., Zolghadri, M. R., & Alavi, S. M. H. (2012). A design procedure for optimizing the LLC resonant converter as a wide output range voltage source. *IEEE Transactions on Power Electronics, 27*(8), 3749–3763.
5. Chandgude, S., & Salunkhe, S. (2021). In state of art: Mechanical behavior of natural fiber-based hybrid polymeric composites for application of automobile components. *Polymer Composites, 42*(6), 2678–2703.
6. Conrad, H. (2000). Influence of an electric or magnetic field on the liquid–Solid transformation in materials and on the microstructure of the solid. *Materials Science and Engineering: A, 287*(2), 205–212.
7. Hoult, D. I., & Tomanek, B. (2002). Use of mutually inductive coupling in probe design. *Concepts in Magnetic Resonance: An Educational Journal, 15*(4), 262–285.
8. Jang, Y. J., Jeong, S., & Ko, Y. D. (2015). System optimization of the On-Line Electric Vehicle operating in a closed environment. *Computers & Industrial Engineering, 80*, 222–235.
9. Johansson, O. (2009). Disturbance of the immune system by electromagnetic fields—A potentially underlying cause for cellular damage and tissue repair reduction which could lead to disease and impairment. *Pathophysiology, 16*(2–3), 157–177.
10. Kheifets, L., Repacholi, M., Saunders, R., & Van Deventer, E. (2005). The sensitivity of children to electromagnetic fields. *Pediatrics, 116*(2), e303–e313.
11. Ko, Y. D., & Jang, Y. J. (2013). The optimal system design of the online electric vehicle utilizing wireless power transmission technology. *IEEE Transactions on Intelligent Transportation Systems, 14*(3), 1255–1265.
12. Koch, C. C. (2000). Experimental evidence for magnetic or electric field effects on phase transformations. *Materials Science and Engineering: A, 287*(2), 213–218.
13. Kuipers, J., Bruning, H., Bakker, S., & Rijnaarts, H. (2012). Near field resonant inductive coupling to power electronic devices dispersed in water. *Sensors and Actuators A: Physical, 178*, 217–222.
14. Lee, S., Choi, B., & Rim, C. T. (2013). Dynamics characterization of the

inductive power transfer system for online electric vehicles by Laplace phasor transform. *IEEE Transactions on Power Electronics, 28*(12), 5902–5909.

15. Li, S., & Mi, C. C. (2014). Wireless power transfer for electric vehicle applications. *IEEE Journal of Emerging and Selected Topics in Power Electronics, 3*(1), 4–17.
16. Liu, L., Liu, J., Gao, B., Yang, F., & Chellam, S. (2012). Fouling reductions in a membrane bioreactor using an intermittent electric field and cathodic membrane modified by vapor phase polymerized pyrrole. *Journal of Membrane Science, 394*, 202–208.
17. Lottermoser, T., Lonkai, T., Amann, U., Hohlwein, D., Ihringer, J., & Fiebig, M. (2004). Magnetic phase control by an electric field. *Nature, 430*(6999), 541–544.
18. Lu, F., Zhang, H., & Mi, C. (2017). A review on the recent development of capacitive wireless power transfer technology. *Energies, 10*(11), 1752.
19. Magoni, D., & Pansiot, J. J. (2001). Analysis of the autonomous system network topology. *ACM SIGCOMM Computer Communication Review, 31*(3), 26–37.
20. Mahadevan, P., Krioukov, D., Fall, K., & Vahdat, A. (2006). Systematic topology analysis and generation using degree correlations. *ACM SIGCOMM Computer Communication Review, 36*(4), 135–146.
21. Markov, M. S. (2007). Expanding use of pulsed electromagnetic field therapies. *Electromagnetic Biology and Medicine, 26*(3), 257–274.
22. Matsukura, F., Tokura, Y., & Ohno, H. (2015). Control of magnetism by electric fields. *Nature Nanotechnology, 10*(3), 209–220.
23. Mohamed, A. A., An, S., & Mohammed, O. (2018). Coil design optimization of power pad in IPT system for electric vehicle applications. *IEEE Transactions on Magnetics, 54*(4), 1–5.
24. Mou, X., Gladwin, D. T., Zhao, R., & Sun, H. (2019). Survey on magnetic resonant coupling wireless power transfer technology for electric vehicle charging. *IET Power Electronics, 12*(12), 3005–3020.
25. Oyama, H., Shinzato, T., Hayashi, K., Kitajima, K., Ariyoshi, T., & Sawai, T. (2008). Application of superconductors for automobiles. *SEI Tech. Rev, 67*, 22–26.
26. Padmanabhan, V., & Rao, R. C. (1993). Warranty policy and extended service contracts: Theory and an application to automobiles. *Marketing Science, 12*(3), 230–247.
27. Park, C., Lee, S., Jeong, S. Y., Cho, G. H., & Rim, C. T. (2015). Uniform

power I-type inductive power transfer system with DQ-power supply rails for on-line electric vehicles. *IEEE Transactions on Power Electronics, 30*(11), 6446–6455.

28. Phadke, M. S., & Dehnad, K. (1988). Optimization of product and process design for quality and cost. *Quality and Reliability Engineering International, 4*(2), 105–112.
29. Schoenbach, K. H., Peterkin, F. E., Alden, R. W., & Beebe, S. J. (1997). The effect of pulsed electric fields on biological cells: Experiments and applications. *IEEE Transactions on Plasma Science, 25*(2), 284–292.
30. Shinohara, N. (2012). The wireless power transmission: Inductive coupling, radio wave, and resonance coupling. *Wiley Interdisciplinary Reviews: Energy and Environment, 1*(3), 337–346.
31. Shoki, H. (2013). Issues and initiatives for practical deployment of wireless power transfer technologies in Japan. *Proceedings of the IEEE, 101*(6), 1312–1320.
32. Sobieszczanski-Sobieski, J., & Haftka, R. T. (1997). Multidisciplinary aerospace design optimization: Survey of recent developments. *Structural Optimization, 14*, 1–23.
33. Sundaresan, S., Ishii, K., & Houser, D. R. (1995). A robust optimization procedure with variations on design variables and constraints. *Engineering Optimization, 24*(2), 101–117.
34. Tao, R., & Tang, H. (2014). Reducing viscosity of paraffin base crude oil with electric field for oil production and transportation. *Fuel, 118*, 69–72.
35. Triviño, A., González-González, J. M., & Aguado, J. A. (2021). Wireless power transfer technologies applied to electric vehicles: A review. *Energies, 14*(6), 1547.
36. Wang, Q., Li, Y., Sun, D. W., & Zhu, Z. (2018). Enhancing food processing by pulsed and high voltage electric fields: Principles and applications. *Critical Reviews in Food Science and Nutrition, 58*(13), 2285–2298.
37. Weisheit, M., Fähler, S., Marty, A., Souche, Y., Poinsignon, C., & Givord, D. (2007). Electric field-induced modification of magnetism in thin-film ferromagnets. *Science, 315*(5810), 349–351.
38. Xanthakis, E., Havet, M., Chevallier, S., Abadie, J., & Le-Bail, A. (2013). Effect of static electric field on ice crystal size reduction during freezing of pork meat. *Innovative Food Science & Emerging Technologies, 20*, 115–120.
39. Young, L. C., & Finlayson, B. A. (1976). Mathematical models of the monolith catalytic converter: Part II. Application to automobile exhaust. *Chemical Engineering Science, 22*(2), 343–353.

40. Yu, X., Du, S., Xu, Z., He, J., Liu, F., Wang, B., & Zhao, K. (2024). Surface-enhanced bimetallic effect of Au-Pd by internal electromagnetic fields from Au@ Cu_2O for efficient electrochemical nitrate reduction to ammonia. *Chemical Engineering Journal, 480*, 148152.
41. Zakian, P., & Kaveh, A. (2024). Multi-objective seismic design optimization of structures: A review. *Archives of Computational Methods in Engineering, 31*(2), 579–594.
42. Zhang, X., Yuan, Z., Yang, Q., Li, Y., Zhu, J., & Li, Y. (2016). Coil design and efficiency analysis for dynamic wireless charging system for electric vehicles. *IEEE Transactions on Magnetics, 52*(7), 1–4.

CHAPTER 8

WPT IN ELECTRIC VEHICLES: EVALUATION, REGULATION, AND FUTURE PROSPECTS

LEARNING OBJECTIVES

At the end of this chapter, readers will be able to:

1. Understand the historical evolution of Electric Vehicles (EVs) and charging methods.
2. Know the different types of conductive charging methods for EVs.
3. Explore the development and standardization of Wireless Power Transfer (WPT) technology for EVs.
4. Understand the field evaluations of WPT systems in various regions like Europe, Japan, Korea, and China.
5. Explore the regulatory and standardization efforts for WPT in EVs.
6. Explore the prospects of EV Wireless Power Transfer Technologies, including efficiency, transmitted power, and reliability, interoperability, and ecosystem enhancement.

INTRODUCTORY EXAMPLE

EcoPower Mobility Solutions is at the forefront of the sustainable transportation industry, thanks to its groundbreaking WPT (Wireless Power Transfer) technologies. The company's commitment to transforming EV technology is clear in its creation of modern WPT systems. Urban environments are greatly benefited by the seamless integration of EcoPower's EVs into daily life, providing a more environmentally friendly transportation option.

The EcoDrive E1, developed by EcoPower, showcases the highest level of engineering EVs. The EcoDrive E1, powered by EcoPower's cutting-edge WPT technology, is a groundbreaking example of the next generation of eco-friendly transportation. In bustling urban areas, the EcoDrive E1 smoothly moves towards designated charging zones, easily recognizable by the stylish charging pads seamlessly integrated towards the pavement.

EcoPower's WPT technology helps the EcoDrive E1 to connect to the charging pads effortlessly, allowing for a smooth and efficient power transfer without the hassle of physical connectors. The vehicle quickly starts recharging the battery, minimizing disturbances to the metropolis, and promoting sustainability.

The WPT system developed by EcoPower streamlines the charging process, guaranteeing exceptional efficiency and reliability. Every time a vehicle is charged, it helps to decrease carbon emissions, promoting sustainability for future generations.

UNIT INTRODUCTION

The possibility of implementing WPT on EVs was first demonstrated by the MIT Team in 2007 and showcased the ability to transfer power over varying air gaps. That is referred to as "MIT's revolution or Columbus's egg (2007)." Following the MIT report, numerous electric vehicle prototypes equipped with WPT capability have been demonstrated and showcased in different automotive exhibitions. During the Tokyo Motor Show 2011 in Japan, various Original Equipment Manufacturers (OEMs) including Toyota, Yamaha, GM, Mitsubishi Motors, Nissan, and showcased their concept EVs equipped with WPT technology. Nissan announced excitingly in the upcoming Tokyo Motor Show 2011 and New York International Auto Show 2012. Their plans were revealed to introduce the Infiniti LE Concept, equipped with a convenient wireless charging option, within the next 24 months. Undoubtedly, this was a groundbreaking announcement to implement a WPT system on passenger EVs produced on a large scale. Unluckily, the plan did not come to fruition (Bi et al., 2016).

At the Paris Motor Show (PMS) on 11th of October 2016, Qualcomm Halo announced the wireless electric vehicle charging technology in the 2018 Mercedes-Benz S550e. This technology was developed using Qualcomm Halo's breakthroughs, as mentioned in their news release. The upcoming update of the Mercedes-Benz S550e in 2018 will include the introduction of a Wireless Electric Vehicle Charging (WEVC) system. This system will be produced by a leading power electronics supplier and will incorporate licensed Qualcomm Halo technology. The use of WEVC in hybrid vehicles is a groundbreaking development in the commercial sector. Owners of the S550e (S500 in Europe) using the wireless charging feature can conveniently park on a designated pad to initiate the charging process. No need to deal with cables or untangling, simply park and charge. Shown in Figure 8.1 is a depiction of a WPT system designed for EVs at a private parking spot (Mahesh et al., 2021). In 2018, Daimler plans to introduce the world's first commercial passenger EV using a wireless charging option, allowing users to easily charge and drive their vehicles. This represents a significant advancement for electric vehicles, offering a more convenient and streamlined charging process through automatic charging, eliminating the need for manual connector operation (Zhang et al., 2018).

Figure 8.1. *Representation of the concept of the WPT system for passenger EV.*

Source: Naoki Shinora; Creative Commons License.

Before this new feature for widespread production of EVs, it is essential to create standardized regulations both internationally and domestically. Additionally, conducting numerous field evaluation tests is necessary. Also, there is a requirement for modifications to domestic and international regulations, including organizations such as the Comite´ International Spe´cial des Perturbations Radioe´lectriques (CISPR) and International Telecommunication Union (ITU). Figure 8.2 presents the comprehensive timeline for standardization of the market beginning, as prepared by the STILL project. The work regarding (STILLE) Standardization of Inductive Charging Systems in Different Power Classes in Germany began when the TIR of SAE J2954 was released in 2016. Simultaneously, the (INL) Idaho National Laboratory in the USA embarked on a comparable endeavor (Hutchinson et al., 2019).

STILLE in Germany is currently exploring the possibility of producing Gen1 private WPT charging by the end of 2017. Additionally, plans are made to introduce Gen2 public WPT charging to the market by the end of 2019 (Vishnuram & Nastasi, 2023).

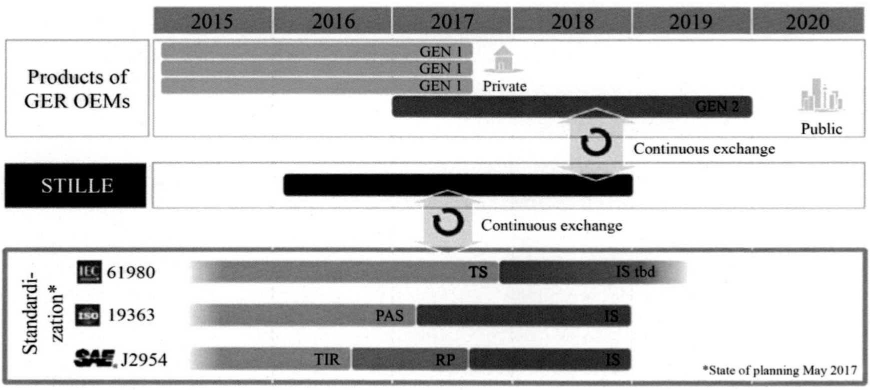

Figure 8.2. *Illustration of overall timeline for standardization with respect to market introduction.*

Source: Naoki Shinora; Creative Commons License.

8.1. EV AND CHARGING

Learning Objectives:

- Understand the historical development and challenges of electric vehicle (EV) charging.

The history of electric vehicles began in 1888 with the Flocken Elektrowagen in Germany. During the New York Motor Show in 1901, there were 28 electric automobiles, 58 gasoline automobiles, and 58 steam automobiles, on display (Figure 8.3).

Due to the challenges with managing charging cables and plugs, and the relatively lower performance of batteries, gasoline-powered vehicles have dominated the transport sector for decades. By the 20th century, the issue of air pollution emerged as a major concern in the realm of transportation, driven by ecological factors. Electric vehicles have been widely recognized as an effective means for reducing CO_2 emissions. Energy storage in electric vehicles often relies on the use of Li-ion batteries and electric double-layer capacitors (Sujitha & Krithiga, 2017).

Various methods have been developed and standardized for charging or transferring energy to an on-vehicle electric storage system. Figure 8.4 displays the categorization of e-charging methods.

To establish a standard for electric vehicle charging that can be widely adopted, various technical committees (TCs) of the International Electrotechnical Commission (IEC) are working together and maintaining close communication with each other, as depicted in Figure 8.5. TC69 focuses on developing global standards over road vehicles or electric industrial trucks that are powered by self-contained power sources. The work on EV WPT systems in TC69, and WG7 began in 2016 (Jian et al., 2018).

Figure 8.3. *Illustration of cable and plug charging for EV at early stage.*

Source: Naoki Shinora; Creative Commons License.

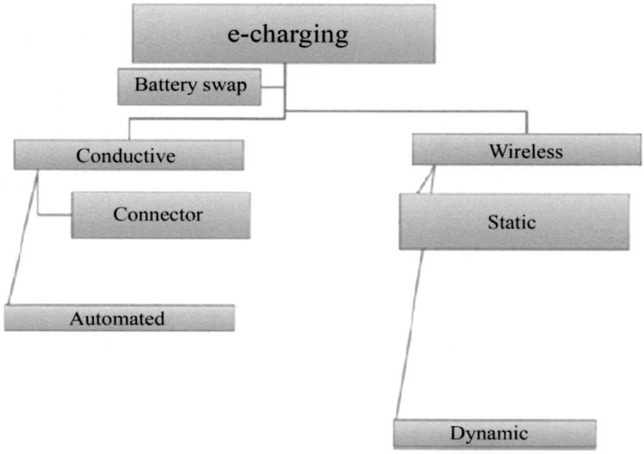

Figure 8.4. *Illustration of classification of e-charging method.*

Source: Kim, K.; Creative Commons License.

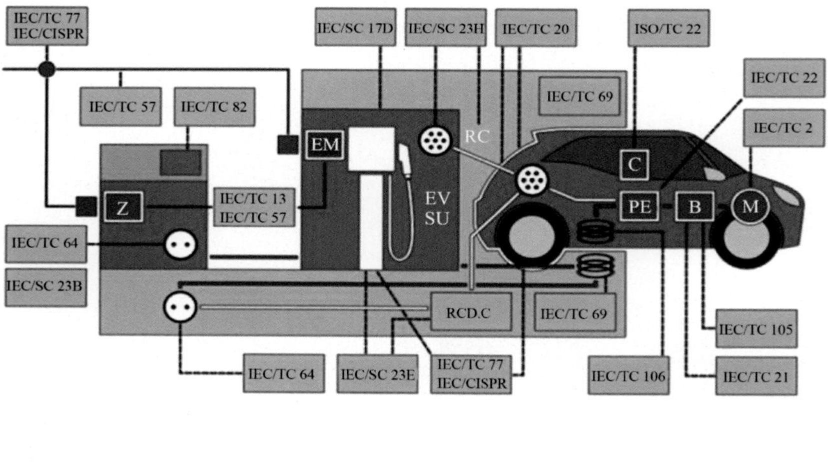

Figure 8.5. *Illustration of technical committees relating e-mobility charging method.*

Abbreviations: B: Battery/fuel cell; C: communication (PWM; V2G); EVSU: electric vehicle supply unit; RC: recharger cable; PE: power electronics; M: (electric) motor; RCD: residual current protective device (circuit breaker); and EM: electricity meter (Yang et al., 2018).

Source: Naoki Shinora; Creative Commons License.

8.2. CONDUCTIVE CHARGING

Learning Objectives:

- Understand the development and deployment of conductive charging methods for EVs.

Figure 8.5 illustrates that numerous TCs of the IEC have established standardization for the conductive charging of EVs. These TCs cover various aspects like cables and connectors (IEC15118 series) and communication (command and control, IEC15118 series). Charging electric vehicles using conductive methods has become a widely used approach, but it will not be the focus of this book. For a comprehensive understanding, refer to Section 8.4 of the Automotive Engineering Handbook Volume 7 (EV/Hybrid) by JSAE (Wang et al., 2013).

Conductive AC charging gained popularity in Europe, leading to the introduction of numerous charging stations in European cities. One example depicted in Figure 8.6 displays an AC conductive charger conveniently located near the NICE central station (Vu et al., 2020).

There is a growing popularity in Europe for fast conductive charge stations that allow for multiple uses of different types of charging, including Combo, ChaDeMo for fast DC charging, and AC charging. These stations help to overcome the challenges posed by different connector standards. Figure 8.7 depicts the various applications of a rapid conductive station for charging at the Vattenfall office in Stockholm, Sweden, in April 2016 (Chun et al., 2010).

Figure 8.6. *Illustration of AC conductive charging spot at NICE central station.*

Source: Brad, B.; Creative Commons License.

Figure 8.7. *The fast conductive charging station of Combo, CHaDeMo, and AC at Vattenfall office of Stockholm, Sweden.*

Source: Naoki Shinora; Creative Commons License.

Both the alternating current and direct current charging techniques require the use of a connector and cable, which must be handled manually. In certain European cities, EV Bus has set up advanced high-power charging systems for EVs, eliminating the need for cables, pantographs, or other traditional methods. In August 2015, Volvo bus in Goteborg, Sweden, had already implemented automatic high-power charging stations at the terminal and bus stop (Zhao et al., 2005). The required charging time is a mere 6 minutes, as shown in Figure 8.8.

Figure 8.8. *Illustration of automated high-power charging station for Volvo bus of Goteborg, Sweden.*

Source: Naoki Shinora; Creative Commons License.

8.3. WIRELESS CHARGING

Learning Objectives:

- Understand the global field evaluations of Wireless Power Transfer (WPT) for EVs.
- Examine the development and standardization efforts of WPT technology in different countries.

The standardization and commercialization of WPT for EVs began in California, USA, towards the end of 2000. The SAE J1773 standard for Inductively Coupled Charging in Electric Vehicles was published in 1995 (Hui, 2013).

Figure 8.9 illustrates the charging style of the GM EV-1, while Figure 8.10 displays the charging paddle. This particular paddle design allows for a minimal distance between the primary and secondary coil, without the need for a physical connection through wires. It is commonly used in the field of inductive charging for electric vehicles (Chittoor et al., 2021).

8.3.1. Field Evaluation in Europe

Following the breakthrough of "MIT's revolution or Columbus's egg" in 2007, the issue of air gap restriction in WPT for EVs was successfully resolved. Several studies and field evaluations have suggested deploying on passenger electric vehicles (Lühken et al., 2014).

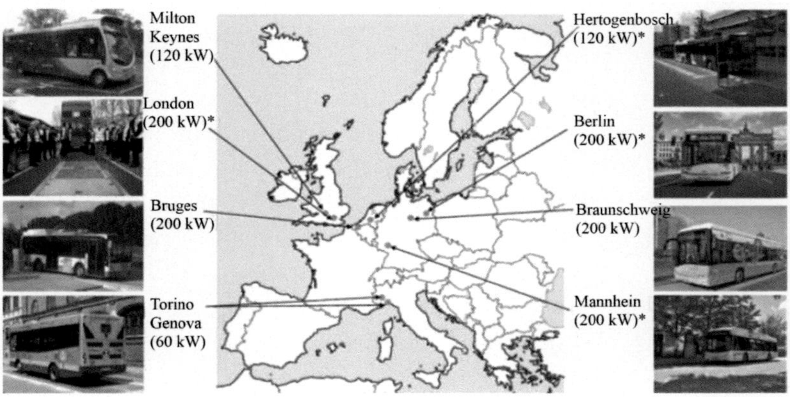

Figure 8.9. *High-power wireless charging application for public service in Europe.*

Source: Daniel, P.; Creative Commons License.

On November 18, 2011, Qualcomm Halo made one of the earliest evaluations. In a recent press release, Qualcomm Incorporated proudly announced the launch of the first trial for Wireless Electric Vehicle Charging (WEVC) in London. This groundbreaking initiative is set to make waves in both the UK and the industry as a whole. Qualcomm is working together with the U.K. Government, the Mayor of London's office, and Transport for London (TfL) to conduct the trial. They also stated that a pre-commercial trial is anticipated to commence in early 2012 and will include up to 50 EVs. Regrettably, a report on the pre-commercial trial has not been located at this time (Targetti et al., 2014).

In Europe, the operation of electric buses for public transport has been extensively evaluated and is now being used in some commercial operations. Figure 8.9 depicts the field evaluation of WPT implemented on an EV bus. This particular system was developed and introduced by Takahashi.

Practice Problem

ElectroDrive Technologies, a pioneering company in the Electric Vehicle (EV) industry, is conducting a field evaluation of their latest Wireless Power Transfer (WPT) system in Europe. They find that the system reliably transmits 8 kW of power wirelessly to participating EVs. However, the system requires an input power of 10 kW from the grid to achieve this. Calculate the efficiency of ElectroDrive's WPT system during the evaluation.

Solutions to Practice Problem

During the field evaluation in Europe, ElectroDrive Technologies determined that their WPT system transmitted 8 kW of power wirelessly to EVs while drawing 10 kW of input power from the grid. To calculate the efficiency of their system, we use the formula: Efficiency (%) = (Output Power/Input Power) × 100. Substituting the given values, the efficiency of ElectroDrive's WPT system can be calculated as follows: Efficiency (%) = (8 kW/10 kW) × 100 = 80%. Therefore, during the field evaluation, ElectroDrive's WPT system exhibited an efficiency of 80%, indicating that 80% of the input power from the grid was effectively transferred to the EVs wirelessly.

8.3.2. Field Evaluation in Japan

In February 2012, Mitsui Home and IHI disclosed their plans to evaluate WPT technology in EVs as part of the Smart Home project HEMS in Kashiwanoha, Chiba Prefecture, Japan. An evaluation was conducted in March 2016, utilizing a smartphone for monitoring, and controlling the charging of WPT in electric vehicles. In Japan, both Toyota and Honda conducted field evaluations of wireless charging for electric vehicles in 2014 (Matsuda, 2016). Toyota conducted an evaluation of three electric vehicles over two years, including their use in residential settings. Honda collaborates with the government of Saitama town on the innovative smart home project called "E-KIZUNA Project." Toshiba, in collaboration with Waseda University and ANA, did an assessment of wireless charging for both electric vehicles and buses over two years. This research was made possible through funding from a Ministry of the Environment project (Iwasaki et al., 2013).

8.3.3. Field Evaluation in Korea

In Korea, the KOREAN Wireless Power Forum (KWPF) was established in 2011. They have alliances with domestic companies and Standards Development Organizations (SDOs) (Jones et al., 2013). Unfortunately, the forum nearly disregarded the concept of WPT charging for EVs. Research and practical assessment on electric buses in Korea were conducted by the Korean Advanced Institute of Science and Technology (KAIST). Research is being conducted on the study and development of dynamic as well as static wireless charging for an electric vehicle bus known as the "online electric vehicle (OLEV)." Figures 8.10 and 8.11 depict the operation of OLEV on the campus at Daejeon, Korea (Lee et al., 2014).

Figure 8.10. *Illustration of wireless charging bus and secondary coils by Toshiba.*

Source: Daniel, P.; Creative Commons License.

Figure 8.11. *Illustration of OLEV at KAIST campus, Daejeon, Korea.*

Source: Hustler Jo; Creative Commons License.

8.3.4. Field Evaluation in China

China has been researching and implementing WPT for electric buses and passenger EVs. The initial report on WPT for EV was showcased at the UNECE/WP29, EVE 7th session in Beijing, China on October 17–18, 2013. Through studies for 5 years, ZTE successfully developed WPT for buses and EVs. And formed a strategic partnership in 38 cities in mainland China. ZTE China has made significant advancements in the field of EV charging, both in terms of conductive and wireless charging technologies. These developments are specifically tailored for EVs and electric buses (Chen et al., 2009).

ZTE is actively working towards standardizing EV WPT in China, as mentioned in Article 15. ZTE has provided sponsorship to the Wireless Power Transmission Alliance, an industry alliance focused on the development of WPT technology, the creation of group standards, and the promotion of the market. They also advocate for the creation of a collaborative working group between the National Technical Committee of Automotive Standardization (NTCAS), and China Electricity Council (CEC). This group would be responsible for standardizing the GB/QC/NB standards for Electric Vehicle WPT. ZTE is actively encouraging the adoption of local standards for WPT in electric vehicles across various provinces and cities (Hu et al., 2019).

8.4. REGULATION AND STANDARDIZATION FOR WPT

Learning Objectives:

- Understand the regulatory considerations and standardization efforts for wireless power transfer (WPT).

To successfully implement and bring WPT for EVs into the market, it is essential to establish regulations that pertain to wireless services. Ensuring the safety of individuals and minimizing any disruption to current wireless services are both crucial considerations. It is essential to mandate domestic and international regulations overseen by local government. The Ministry of Information and Communication (MIC) in Japan, the Federal Communications Commission (FCC) in the United States, the European Telecommunications Standards Institute (ETSI), and the Comite´ Europe´en de Normalization E´lectrotechnique (CENELEC) in Europe are actively engaged in regulatory work (Shoki, 2013). It is essential to give careful consideration to standardizing and regulation for passenger electric vehicles, as depicted in Figure 8.12.

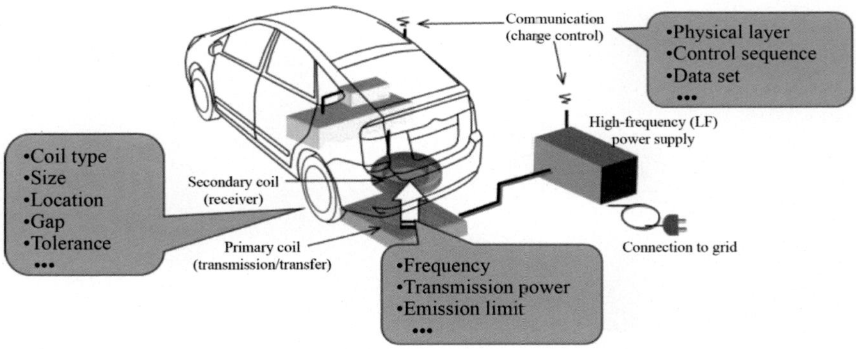

Figure 8.12. *Illustration of standardization part for EV WPT.*

Source: Daniel, P.; Creative Commons License.

1. **Interoperability:** Compatible with various original equipment manufacturers (OEMs), power classes, height classes, and other specifications (Koohestani et al., 2016).
2. **Safety:** It is important to take into account human safety, as outlined by the International Commission on Non-Ionizing Radiation Protection (ICNIRP). Additionally, it is important to consider the

impact of active implantable medical devices (AIMD) under the EU directive (90/385/EEC). It is crucial to prioritize the implementation of foreign object detection systems to prevent fires caused by excessive temperature rise.

3. In coexistence with other existing services, it is important to take into account the WPT of the magnetic coupling method and its utilization of frequency resources. This will ensure compliance with radioactivity and frequency management regulations (Hui, 2018).

It is important to take into account the potential impact on existing radio services, both in terms of the frequency used for WPT and any out-of-band frequencies that can interact. International and domestic standardization and regulation organizations have a close relationship with each other. Figure 8.13 illustrates the relationship (Hossain et al., 2022).

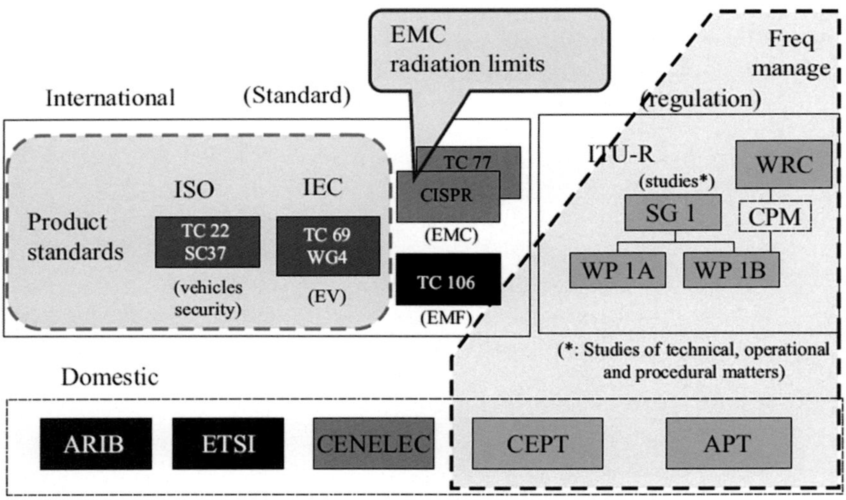

Figure 8.13. *Illustration of WPT relating international and domestic organizations.*

Source: Xi Zhang, Creative Commons License.

8.4.1. European Standards for Electricity Supply

Following discussions held by IEC TC69 WG4 and ISO TC22 SC37, the European Commission issued Commission Implementing Decision M/533 on 3 December 2015. According to the document, on page 7, it mentions a technical solution for ensuring interoperability in wireless recharging passenger cars and light-duty vehicles. This solution is compatible with the specifications outlined in IEC/TS 61980-3 Ed. 1.0 or any subsequent editions. As per the M/533 request, European OEMs and Tier 1 supply companies, formed STILLE to provide their input to IEC TC69 WG7, ISO TC22, SC37, and SAE. The organizations IEC/ISO and SAE are involved in this process (Pérez, 2007).

8.4.2. IEC/ISO and SAE

Figure 8.14 displays the standardized WPT system (ISO 19363/IEC 61851-3/SAE J2954) during its initial phase.

Discussion on the WPT system for passenger EVs is centered around several important factors:

- The WPT system is designed for passenger cars and light-duty vehicles.
- The WPT system utilizes a magnetic field (MF-WPT).
- The main device (on the infrastructure side ¼ supply side) is installed upon the surface level.
- The secondary device is installed below the vehicle's underbody (Rezaei et al., 2014).
- The frequency range of 85 kHz (79.00–90.00 kHz) has been suggested and established as the designated band for power transfer in ISO/PAS 19363:2016, SAE/TIR J2954:2016, and IEC 61980-3 Committee Draft.
- Technical information reports (TIR) and publicly available specifications (PAS) are important resources in the field of EVs:
- Standardization efforts are currently underway to establish standards for power classes < 3.7 kW 3.7–7.7 kW, and 7.7–11 kW in EVs.

The discussion on the WPT system over heavy-duty vehicles was initiated in the SAE J2954 TF. There is a classification system established by IEC, ISO, or SAE that categorizes input power for powering devices in electric vehicles (Green et al., 2016).

Ensuring interoperability among various primary and secondary coil types, such as circular or Double D-shaped (DD) types, is crucial for maintaining high efficiency in WPT operations. This is especially important when considering different power classes and Z-class requirements. Figure 8.15 displays the interoperability test conditions for various coils conducted by STILLE (Ko & Sung, 2017).

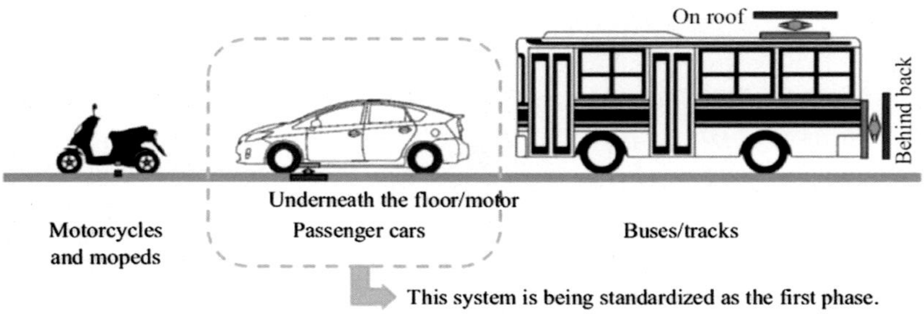

Figure 8.14. *Illustration of system being standardized as first phase.*

Source: Xi Zhang; Creative Commons License.

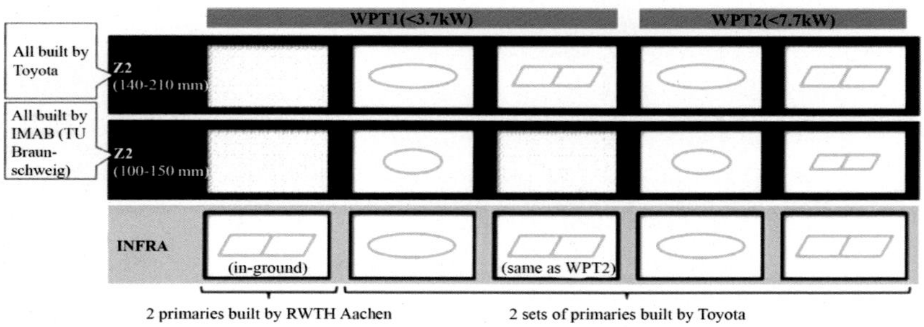

Figure 8.15. *Illustration of interoperability test condition for different coils by STILLE.*

Source: Xi Zhang; Creative Commons License.

8.5. FUTURE OF EV WIRELESS POWER TRANSFER TECHNOLOGIES

Learning Objectives:

- Understand the key trends in Wireless Power Transfer (WPT) technologies for EVs.
- Identify challenges and solutions for improving WPT efficiency, power, reliability, and interoperability.

Since the commercialization of WPT in the 1980s, there has been a significant surge in its development in recent years. After three decades, with the pressing issues of resource scarcity and environmental degradation, there has been a remarkable surge of interest in WPT (Imura et al., 2016). Its development holds great potential in addressing this complex challenge. The future is looking bright, with a summary of the latest trends in WPT development presented in Figure 8.16. Further details are discussed in subsections.

8.5.1. High Efficiency

WPT is a device that transmits energy, and the efficiency of the system is a fundamental factor in describing its performance. Typically, efficiency is a key consideration in the realm of electric vehicles. As per the widely accepted definition, the measurement of system efficiency is based on the ratio of the power output of the dc-load to the DC-input power at the inverter. Improved system performance is a key focus in the advancement of WPT technology. On the other hand, when compared with plug-in charging, the WPT system exhibits a slightly lower system efficiency of around 5% to 6%. When the distance between the onboard receiving coil (secondary side coil) and the ground-transmitting coil (primary side coil) is far or offset, the system efficiency tends to decrease. Maintaining high efficiency, both in alignment and misalignment, will continue to be a significant challenge for the foreseeable future (Adachi et al., 2001).

8.5.2. High Transmitted Power

Transmitted power capability is a crucial evaluation criterion for WPT. When the transmitted power is lower, it results in longer charging times for the equipment. This means that consumers have to spend longer on charging, which unfortunately does not help alleviate range anxiety. However, the limited transmitted power has hindered the widespread adoption of WPT. There is still significant potential for improvement in achieving higher transmitted power (Correia et al., 2007).

Exploring the potential of new electronic components could provide a solution to the challenges posed by the resonant inverter. A new inverter was proposed to enhance the power ability of the system, within the MHz frequency band. The inverter utilizes gallium nitride (eGaN) technology. Although the single-phase inverter is useful, the multiphase parallel resonate inverter can also be advantageous in achieving a high-power level (Steigenberger et al., 2018).

Figure 8.16. *Illustration of trends of E-wireless power transfer technologies.*

Source: Chong Zhu; Creative Commons License.

8.5.3. System Reliability

Ensuring the utmost validity of WPT is of utmost importance, given the intricate nature of the system or the ever-changing environment it operates. In the previous section, it was discussed that the system consists of various components. If any of these parts malfunction, it can result in the entire system losing control, breaking down, and creating unforeseen safety risks. Due to the increasing power transmission, certain components of the wireless power transfer (WPT) system are prone to experiencing failures. Furthermore, the technology is not limited to indoor use, as it is also utilized in dynamic charging scenarios that take place outdoors. It is important to consider the impact of a changeable environment on the reliability of systems (Kuo & Prasad, 2000).

8.5.4. Interoperability

Establishing a universal standard for devices is challenging due to the unique nature of the WPT system. Using the WPT system in electric vehicles as an illustration, it is worth noting that the battery capacity and placement vary among different models and manufacturers. This aspect becomes particularly crucial when creating a WPT system. Table 8.1 presents the power battery characteristics of three commercially available electric vehicles (Wegner, 1996).

Table 8.1. *Illustration of Power Battery Parameters of Four Commercial EVs*

Model	Rated Voltage of Battery Pack (V)	Battery Pack Capacity (kWh)	Pure Electric Mileage (km)
Tesla model S	360	100	570
Toyota PRIUS (plug-in)	346	4.4	22
BMW mini-E	380	35	250

Source: Chong Zhu; Creative Commons License.

When it comes to the widespread use of WPT, it is crucial to take into account the compatibility of various coils and topologies, power levels, and models from different manufacturers (Heiler, 1995).

Since the introduction of wireless charging technology, three distinct standards organizations have emerged, namely A4WP, PMA, and WPC, proposing QI standards. The A4WP standard utilizes resonant technology, whereas PMA and QI rely on inductive charging. It can be challenging to achieve compatibility among the three different standards. Furthermore, the lack of compatibility poses a challenge to the advancement of WPT to some degree (Zeng, 2019).

To expedite the widespread adoption of WPT, it is crucial to consolidate the efforts of all three groups and establish a comprehensive global standard that can be applied to all devices. In 2014, A4WP made a significant move by merging with PMA. Undoubtedly, this is a driving force behind the advancement of wireless charging. Certainly, the compatibility of various standards is an inevitable trend in WPT. Nevertheless, there is a certain amount of time involved (Litwin & Abdellatif, 1986).

8.5.5. Improvement of Ecosystem

WPT is a crucial component that should be integrated into the construction of an ecosystem for intelligent EVs. It is not an alone technology, but rather a fundamental technology that plays a vital role in the development of EVs. Furthermore, Figure 8.17 illustrates the arrangement of the ecosystem (Xu, 2019). As previously stated, the accuracy of positioning has a significant impact on the performance of WPT. Another option for enhancing misalignment capacity is to integrate the system using driverless and automatic parking technology (Xu et al., 2023).

One could refer to the combination as a comprehensive network of intelligent electric vehicles. All three methods work in synergy, with the high-precision positioning of driverless and automatic parking enhancing the efficiency of the WPT system, whereas the WPT system enables driverless and automatic parking (Gontareva et al., 2022).

In addition, Internet companies are committed to designing their products with a focus on the most widely used personal mobile device, automobiles. Additionally, the presence of well-developed supporting facilities is crucial for the overall ecosystem, including the inclusion of a wireless charging parking lot (Zhai et al., 2020).

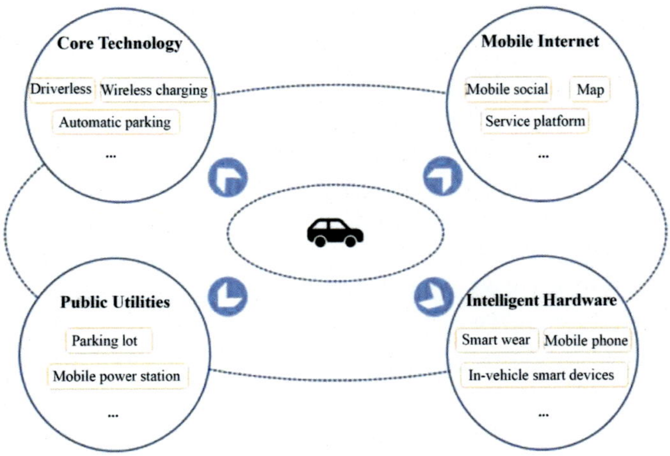

Figure 8.17. *Illustration of ecosystem of intelligent EVs.*

Source: Haitao Song; Creative Commons License.

SUMMARY

- The chapter on Wireless Power Transfer (WPT) in Electric Vehicles (EVs) explores the multifaceted landscape of EV charging, with a particular focus on wireless charging technologies. It delves into both conductive and wireless charging methods, highlighting the advantages and challenges of each approach. T
- The chapter presents field evaluations conducted in Europe, Japan, Korea, and China, providing insights into the practical implementation and performance of WPT systems across different regions.
- Regulation and standardization play a crucial role in the widespread adoption of WPT technologies. The chapter discusses existing standards and regulations governing WPT, including European standards for electricity supply and standards set forth by international bodies such as IEC/ISO and SAE.
- Looking ahead, the chapter outlines the prospects of EV WPT technologies, emphasizing the need for advancements in efficiency, power transmission capabilities, system reliability, and interoperability.
- These developments are essential for driving the continued growth and maturation of the EV market, as well as for enhancing the overall user experience and expanding the accessibility of electric transportation.

REVIEW QUESTIONS

1. Contrast Wireless Power Transfer (WPT) with traditional conductive charging for EVs.
2. Enumerate advantages of Wireless Power Transfer (WPT) in EVs.
3. Summarize field evaluations of WPT in Europe, Japan, Korea, and China.
4. Outline current regulations and standardization for WPT in EVs across regions.
5. Explain the significance of European Standards for WPT in EVs.
6. Compare IEC/ISO and SAE standards for WPT in EVs.
7. Discuss future prospects of EV WPT considering efficiency, power, reliability, interoperability, and ecosystem.
8. Address challenges of interoperability in WPT for EVs.

9. Analyze potential impacts of WPT advancements on the automotive industry and society.
10. Identify key research areas to enhance WPT adoption in EVs.

MULTIPLE CHOICE QUESTIONS

1. What does WPT stand for in the context of Electric Vehicles?

a. Wireless Power Transmission

b. Wired Power Transfer

c. Wide Performance Testing

d. Wattage Power Technology

2. Among the listed regions, which one has NOT conducted field evaluations of WPT for EVs as mentioned in the chapter?

a. Europe

b. North America

c. Japan

d. Korea

3. What role do regulations and standardization play in the adoption of WPT for EVs?

a. They ensure wireless internet connectivity.

b. They guarantee high-speed charging.

c. They facilitate safety, compatibility, and interoperability.

d. They provide guidelines for road usage.

4. Which international body is responsible for setting standards for WPT?

a. WHO

b. IEC/ISO

c. NASA

d. CIA

5. What is identified as a key area for improvement in the future of EV WPT technologies?

a. Decreased power transmission capabilities

b. Reduced system reliability

c. Higher efficiency

d. Decreased interoperability

6. **Which of the following is NOT identified as a challenge in the implementation of WPT for EVs?**
   a. System reliability
   b. Power transmission capabilities
   c. Standardization
   d. Interoperability

Answers to Multiple Questions

1. (a); 2. (b); 3. (c); 4. (b); 5. (c) 6. (c)

REFERENCES

1. Adachi, C., Baldo, M. A., Forrest, S. R., Lamansky, S., Thompson, M. E., & Kwong, R. C. (2001). High-efficiency red electro phosphorescence devices. *Applied Physics Letters, 78*(11), 1622–1624.
2. Bi, Z., Kan, T., Mi, C. C., Zhang, Y., Zhao, Z., & Keoleian, G. A. (2016). A review of wireless power transfer for electric vehicles: Prospects to enhance sustainable mobility. *Applied Energy, 179*, 413–425.
3. Chen, Y., Yang, Z., Shu, F., Hu, Z., Meyer, M., & Bhattacharya, S. (2009). A patent-based evaluation of technological innovation capability in eight economic regions in PR China. *World Patent Information, 31*(2), 104–110.
4. Chittoor, P. K., Chokkalingam, B., & Mihet-Popa, L. (2021). A review on UAV wireless charging: Fundamentals, applications, charging techniques and standards. *IEEE Access, 9*, 69235–69266.
5. Chun, H., Chung, T. D., & Ramsey, J. M. (2010). High yield sample preconcentration using a highly ion-conductive charge-selective polymer. *Analytical Chemistry, 82*(14), 6287–6292.
6. Correia, L. H., Macedo, D. F., dos Santos, A. L., Loureiro, A. A., & Nogueira, J. M. S. (2007). Transmission power control techniques for wireless sensor networks. *Computer Networks, 51*(17), 4765–4779.
7. Gontareva, I., Litvinov, O., Hrebennyk, N., Nebaba, N., Litvinova, V., & Chimshir, A. (2022). Improvement of the innovative ecosystem at universities. *Eastern-European Journal of Enterprise Technologies, 1*(13), 115.
8. Green, J. M., Hartman, B., & Glowacki, P. F. (2016). A system-based view of the standards and certification landscape for electric vehicles. *World Electric Vehicle Journal, 8*(2), 564–575.

9. Heiler, S. (1995). Semantic interoperability. *ACM Computing Surveys (CSUR), 27*(2), 271–273.
10. Herzog, W. (2014). Mechanisms of enhanced force production in lengthening (eccentric) contractions. *Journal of Applied Physiology, 116*(11), 1407–1417.
11. Hossain, M. S., Kumar, L., El Haj Assad, M., & Alayi, R. (2022). Advancements and future prospects of electric vehicle technologies: A comprehensive review. *Complexity, 2022*, 1–21.
12. Hu, Y., Wang, C., Xiao, G., & Zeng, J. (2019). The value of political connections in opaque firms: Evidence from China's file 18. *Pacific-Basin Finance Journal, 55*, 329–351.
13. Hui, S. R. (2018). Technical and safety challenges in emerging trends of near-field wireless power transfer industrial guidelines. *IEEE Electromagnetic Compatibility Magazine, 7*(1), 78–86.
14. Hui, S. Y. (2013). Planar wireless charging technology for portable electronic products and Qi. *Proceedings of the IEEE, 101*(6), 1290–1301.
15. Hutchinson, L., Waterson, B., Anvari, B., & Naberezhnykh, D. (2019). Potential of wireless power transfer for dynamic charging of electric vehicles. *IEEE Access, 13*(1), 3–12.
16. Imura, T., Yasuda, T., Oshima, K., Nayuki, T., Sato, M., & Oshima, A. (2016). Wireless power transfer for electric vehicle at the kilohertz band. *IEEJ Transactions on Electrical and Electronic Engineering, 11*, S91–S99.
17. Iwasaki, Y., Ozaki, M., Nakamura, K., Horino, H., & Kawashima, S. (2013). Relationship between increment of groundwater level at the beginning of irrigation period and paddy field area in the Tedori River Alluvial Fan Area, Japan. *IEEJ Transactions on Electrical and Electronic Engineering, 11*, 551–558.
18. Jian, L., Yongqiang, Z., & Hyoungmi, K. (2018). The potential and economics of EV smart charging: A case study in Shanghai. *Energy Policy, 119*, 206–214.
19. Jones, C., Lee, C. K., Ahn, C., Shin, J., & Knezevic, I. (2013). Working Group on quality, safety and efficacy of typhoid Vi capsular polysaccharide conjugate, vaccines, Jeju, Republic of Korea, 5–7 September 2012. *Vaccine, 31*(41), 4466–4469.
20. Ko, K., & Sung, D. K. (2017). The effect of cellular network-based communication delays in an EV aggregator's domain on frequency regulation service. *IEEE Transactions on Smart Grid, 10*(1), 65–73.

21. Koohestani, M., Zhadobov, M., & Ettorre, M. (2016). Design methodology of a printed WPT system for HF-band mid-range applications considering human safety regulations. *IEEE Transactions on Microwave Theory and Techniques, 65*(1), 270–279.
22. Kuo, W., & Prasad, V. R. (2000). An annotated overview of system-reliability optimization. *IEEE Transactions on Reliability, 49*(2), 176–187.
23. Lāce, B., & Bankina, B. (2013). Evaluation of European pear rust severity depending on agro-ecological factors. *Research for Rural Development, 1*, 6–12.
24. Lee, Y., Lim, M. C., Son, Y., Joo, J., Park, K., Kim, J. S., & Park, S. Y. (2014). Development and evaluation of Korean version of Quality of Sexual Function (QSF-K) in healthy Korean women. *Journal of Korean Medical Science, 29*(6), 758–765.
25. Litwin, W., & Abdellatif, A. (1986). Multidata base interoperability. *Computer, 19*(12), 10–18.
26. Lühken, R., Pfitzner, W. P., Börstler, J., Garms, R., Huber, K., Schork, N., & Krüger, A. (2014). Field evaluation of four widely used mosquito traps in Central Europe. *Parasites & Vectors, 7*, 1–11.
27. Mahesh, A., Chokkalingam, B., Mihet-Popa, L., Vishnuram, P., P. S., K. V., & Nastasi, B. (2023). Wireless chargers for electric vehicles: A systematic review on converter topologies, environmental assessment, and review policy. *Energies, 16*(4), 1731.
28. Matsuda, S. (2016). Development of case mix based evaluation system in Japan. *Japanese Hospital, 35*, 35–44.
29. Pérez-Arriaga, I. J. (2007). Security of electricity supply in Europe in a short, medium and long-term perspective. *European Review of Energy Markets, 2*(2), 1–28.
30. Rezaei, R., Chiew, T. K., Lee, S. P., & Aliee, Z. S. (2014). Interoperability evaluation models. *Computers in Industry, 65*(1), 1–23.
31. Shoki, H. (2013). Issues and initiatives for practical deployment of wireless power transfer technologies in Japan. *Proceedings of the IEEE, 101*(6), 1312–1320.
32. Steigenberger, P., Thoelert, S., & Montenbruck, O. (2018). GNSS satellite transmit power and its impact on orbit determination. *Journal of Geodesy, 92*(6), 609–624.
33. Sujitha, N., & Krithiga, S. (2017). RES-based EV battery charging system: A review. *Renewable and Sustainable Energy Reviews, 75*, 978–988.

34. Vishnuram, P., P. S., K. V., & Nastasi, B. (2023). Wireless chargers for electric vehicles: A systematic review on converter topologies, environmental assessment, and review policy. *Energies, 16*(4), 1731.

35. Vu, V. B., González-González, J. M., Pickert, V., Dahidah, M., & Triviño, A. (2020). A hybrid charger of conductive and inductive modes for electric vehicles. *IEEE Transactions on Industrial Electronics, 68*(12), 12021–12033.

36. Wang, H., Hasanzadeh, A., & Khaligh, A. (2013). Transportation electrification: Conductive charging of electrified vehicles. *IEEE Electrification Magazine, 1*(2), 46–58.

37. Wegner, P. (1996). Interoperability. *ACM Computing Surveys (CSUR), 28*(1), 285–287.

38. Xu, D. (2019). Research on enterprise performance improvement under the innovative business ecosystem. *Journal of Business Innovation, 3*(1), 26–29.

39. Xu, K., Chen, J., Feng, Y., Wang, J., & Bai, Z. (2023). How are nature-based solutions contributing to the improvement of ecosystem quality in China: A systematic review. *Ecological Indicators, 155*, 110985.

40. Yang, Y., Jia, Q. S., Guan, X., Zhang, X., Qiu, Z., & Deconinck, G. (2018). Decentralized EV-based charging optimization with building integrated wind energy. *IEEE Transactions on Automation Science and Engineering, 16*(3), 1002–1017.

41. Zeng, M. L. (2019). Interoperability. *Knowledge Organization, 46*(2), 122–146.

42. Zhai, T., Wang, J., Jin, Z., Qi, Y., Fang, Y., & Liu, J. (2020). Did improvements of ecosystem services supply-demand imbalance change environmental spatial injustices? *Ecological Indicators, 111*, 106068.

43. Zhang, Z., Zhang, B., Deng, B., Wei, X., & Wang, J. (2018). Opportunities and challenges of metamaterial-based wireless power transfer for electric vehicles. *Wireless Power Transfer, 5*(1), 9–19.

44. Zhao, S., Castle, G. P., & Adamiak, K. (2005). Comparison of conduction and induction charging in liquid spraying. *Journal of Electrostatics, 63*(6–10), 871–876.

INDEX

D

E

F

G

H

I

K

L

M

N

O

P

Q

R

S